Practical Biomechanics for the Orthopedic Surgeon

Second Edition

Practical Biomechanics for the Orthopedic Surgeon

Second Edition

Eric L. Radin, M.D.
Clinical Professor
of Orthopaedic Surgery
Department of Surgery
University of Michigan
Medical School
Ann Arbor, Michigan
Director, Bone and Joint Center
Chairman, Department
of Orthopaedic Surgery
Henry Ford Hospital
Detroit, Michigan

Robert M. Rose, Sc.D., P.Eng.
Professor
Department of Materials Science
and Engineering
Director, Concourse Program
Massachusetts Institute
of Technology
Cambridge, Massachusetts

J. David Blaha, M.D.
Associate Professor
Department of Orthopedic Surgery
West Virginia University
School of Medicine
Chief, Section
of Arthritis Surgery
West Virginia University Hospitals
Morgantown, West Virginia

Alan S. Litsky, M.D., Sc.D.
Assistant Professor
Division of Orthopaedics
Department of Surgery
College of Medicine
and Biomedical Engineering Center
College of Engineering
Director, Orthopaedic
BioMaterials Laboratory
Ohio State University
Columbus, Ohio

Churchill Livingstone
New York, Edinburgh, London, Melbourne, Tokyo

Library of Congress Cataloging-in-Publication Data
Practical biomechanics for the orthopedic surgeon / Eric L. Radin ...
 [et al.]. — 2nd ed.
   p. cm.
  Includes bibliographical references and index.
  ISBN 0-443-08702-4
  1. Orthopedic surgery. 2. Human mechanics. 3. Biomechanics.
I. Radin, Eric L.
  [DNLM: 1. Biomechanics. 2. Orthopedics. WE 103 P895]
RD732.P7 1992
617.3—dc20
DNLM/DLC
for Library of Congress   91-36412
                                                                      CIP

Second Edition© Churchill Livingstone Inc. 1992
First Edition © Churchill Livingstone Inc. 1979
All rights reserved. No part of this publication may be reproduced, stored in a retrieval system, or transmitted in any form or by any means, electronic, mechanical, photocopying, recording, or otherwise, without prior permission of the publisher (Churchill Livingstone Inc., 650 Avenue of the Americas, New York, NY 10011).

Distributed in the United Kingdom by Churchill Livingstone, Robert Stevenson House, 1–3 Baxter's Place, Leith Walk, Edinburgh EH1 3AF, and by associated companies, branches, and representatives throughout the world.

The Publishers have made every effort to trace the copyright holders for borrowed material. If they have inadvertently overlooked any, they will be pleased to make the necessary arrangements at the first opportunity.

Acquisitions Editor: *Robert A. Hurley*
Copy Editor: *Elizabeth Bowman-Schulman*
Production Designer: *Patricia McFadden*
Production Supervisor: *Christina Hippeli*

Printed in the United States of America

First published in 1992   7 6 5 4 3 2 1

Preface to the Second Edition

Since the publication of the first editon of this book in 1979, Newton's laws of mechanics have not changed but the practice of orthopedics has somewhat. We have revised this primer on *Practical Biomechanics for the Orthopedic Surgeon* to take into account several of these changes. We have therefore included: an analysis of some of the more modern spinal fixation systems in chapter 1; some of the newer fracture internal fixation devices in chapter 2; and the concept of microklutziness to the chapter on joints. The biggest change, as the reader might expect, is in the final chapter on joint replacement, the area of clinical orthopedic practice that has changed the most since 1979.

We also had to modify the team that produced the revision, adding Dr. A.S. Litsky and Dr. J.D. Blaha to Drs. Radin and Rose.

We are pleased with the first edition. It was designed to be a concise and useful introductory book to orthopedic biomechanics, and as such, it became an integral part of many orthopedic resident biomechanical teaching programs. We hope that this new edition is even more useful.

We particularly wish to thank Mary Richards, who acted as our editorial associate and made the project very tolerable for all of us.

Eric L. Radin, M.D.
Robert M. Rose, Sc.D., P.Eng.
J. David Blaha, M.D.
Alan S. Litsky, M.D., Sc.D.

Preface to the First Edition

This book is based on a post-graduate course on biomechanics for orthopedic surgeons given under the auspices of Harvard Medical School for the past several years. Our emphasis has always been on the clinical applications of biomechanical principles. We have departed, in this book, from the popular practice of teaching biomechanics by developing the basic principles first and subsequently applying them, as is done in engineering education. Since the audience in this case is medical, the basic principles are developed from clinical examples that are familiar to the student. The relevant engineering principles involved in a clinical problem are fully discussed, even if this entails repeating certain principles from one chapter to the next. We feel such repetition is good pedagogy. Likewise, we have not hesitated to sacrifice rigor for the sake of effective teaching. For a fully rigorous exposition of the principles involved, we can only recommend courses in materials science.

We wish to thank all our students who filled out critiques at our courses. This feedback has in large part directed our efforts. We also wish to thank Robin Lefberg, our illustrator, whose genius for clarifying is graphically apparent in the text; Christine Byda, who started typing drafts of this work several years ago and uncomplainingly followed through to the end; and our wives and children whose family lives were chronically disrupted by this effort.

The Authors

Contents

1	**Biomechanics of the Spine**	**1**
	1 Functional Anatomy	1
	2 Compression and Compression Fractures	2
	3 Mechanics of the Spinal Column - Newton's First and Third Laws	6
	4 Tension	8
	5 Shear	9
	6 Vectors	10
	7 Forces Applied Versus Stresses Developed	13
	8 Bending	15
	9 Stress Concentration-Effect of Disc Degeneration and Spinal Fusion	22
	10 Torsion	24
	11 Lumbosacral Flexion Exercises and Spinal Bracing	25
	12 Spondylolisthesis as a Fatigue Fracture	27
	13 Resistance of the Spine to Bending	28
	14 Mechanics of Straightening a Curved Spine	31
	15 Traction, Casts, and Braces	37
	16 Internal Fixation Devices	44
2	**Mechanics of Fracture and Fracture Fixation**	**53**
	1 Mechanics of Fracture	53
	2 Tensile Stresses in the Long Bones: Bending and Torsion	54
	3 Mechanics of Fracture: Tensile Stress and Stress Concentrations	59
	4 Energetics of Fracture, Fracture Toughness, and Impact	64

	5 *Fatigue Fracture*; "March" Fractures; Resistance of Cortical Bone to Fractures	65
	6 Corrosion of Metallic Implants	70
	7 Implant Materials for Internal Fixation Devices	74
	8 Mechanical Considerations in Treatment of Fractures	82
	9 Internal Fixation Devices: Wire and Tension Bands	85
	10 Internal Fixation Devices: Plates	87
	11 Spiral Fractures	90
	12 Screws	93
	13 Nails, Rods, and Pins	96
	14 Bone Grafting	104
	15 Recapitulation	107

3 Biomechanics of Sports Injuries — 109

1. Sports Injuries and Newton's Third Law — 109
2. Active Musculoskeletal Shock Absorption — 119
3. Sports Injury Resulting from Fatigue — 120
4. Mechanics of Locomotion — 124
5. Mechanics of Muscle Rupture, Bursitis, Tendonitis, and Meniscal Tear — 130

4 Mechanics of Joint Degeneration — 135

1. Osteoarthrosis as a Wear and Tear Phenomenon — 135
2. Stress Distribution within Joints — 136
3. Mechanical Behavior of Articular Cartilage: Viscoelasticity — 138
4. Mechanical Factors in the Wearing Away of Articular Cartilage — 139
5. Mechanical Considerations in the Treatment of Osteoarthrosis — 143
6. Friction Across Joints — 152
7. Attempts to Treat Osteoarthrosis with Artificial Lubricants — 158

5 Joint Replacements — 159

1. General Remarks — 159
2. Forces in the Normal Hip Joint — 161
3. Stress Distribution after Total Hip Replacement — 171
4. Prosthetic Acetabular Stress Distribution and Joint Shear Force — 174
5. Stress in the Acetabular Cement-Bone Interface — 176

6	Stress Distribution in the Femoral Prosthesis	178
7	Stresses in the Femoral Prosthesis Secondary to Fixation	181
8	Forces and Stresses in the Cement and Bone Surrounding the Femoral Prosthesis	183
9	Knee Joint Stability	184
10	Knee Joint Stresses	188
11	Wear	196

Appendix: Explanation of Units   **201**

Glossary   **203**

Index   **209**

Biomechanics of the Spine

The human spinal column is designed to support the upper body through a wide variety of motions while protecting the spinal cord. This combination of functions, protection and flexibility, mandates that spinal motions be controlled and that undesired motions be resisted. This requires a complex anatomic composite of materials and structures capable of withstanding a wide array of positions and forces without loss of stability. Since the spinal column is subjected to many forces, it is a good place to start an overview of musculoskeletal biomechanics.

1 FUNCTIONAL ANATOMY

The function of the spine is to support the upper torso in various positions, providing sufficient flexibility for trunk movement while protecting the spinal cord from injury. Activities of daily living require sophisticated trunk motions such as bending, twisting, and carrying loads. These functions must be performed with extreme stability because spinal dislocation is always disabling and can be catastrophic. Motions of the spine are never frail and "relaxed" but are the result of highly sophisticated bony and soft tissue interactions in concert with active muscle contraction.

The human spine is composed of a series of bone segments connected by discs and ligaments. Flexibility of this rod-like support is achieved by small displacements of its multiple linkages (Fig. 1-1). The advantage of this configuration is that only slight movement of each disc and ligament is necessary for a large, overall range of motion. An inherently stable situation is also created by multiple, slightly movable segments rather than a few highly mobile articulations.

Despite its structural suitability, the spine is prone to mechanical failure. This chapter examines some of these failures, mechanical factors that may

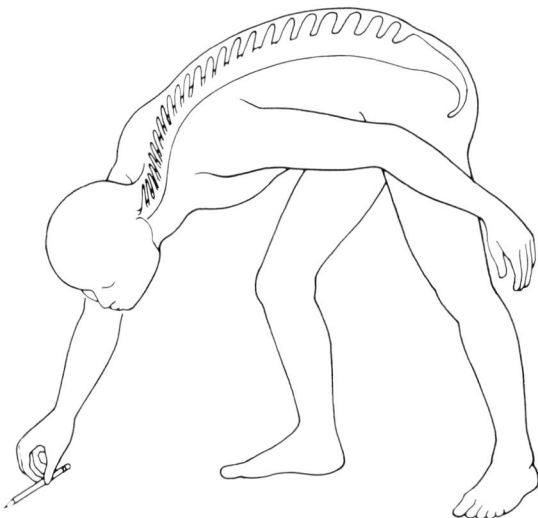

Fig. 1-1. High degree of flexibility achieved in the trunk by small displacements of multiple linkages.

influence their occurrence, and subsequent treatment. Each type of failure is used to introduce biomechanical concepts of general interest and utility.

2 COMPRESSION AND COMPRESSION FRACTURES

Compression* is most simply described as a squeeze. The spine is normally subject to compression because of the weight of the upper body. It is constructed to resist considerable squeezing without crushing. Piled one on top of another, each vertebral body and disc is subjected to compressive force. While being compressed, each unit is deformed by shortening along the direction of compressive force (longitudinal contracture) while expanding laterally (Fig. 1-2). The resistance each has to fracture, its strength, in this case against crushing, depends on two factors: the geometry (size and shape) of the structure and the material(s) from which it is made.

Suppose two identical vertebral bodies are placed side by side. Are they twice as capable of resisting being crushed as one (ie, are they twice as strong)? Intuitively we realize that the strength of each vertebra (in compression) is equal, but now there are two, and the same compressive force is distributed over twice the area (Fig. 1-3). Thus each vertebra receives

---

* Italicized terms are defined in the Glossary.

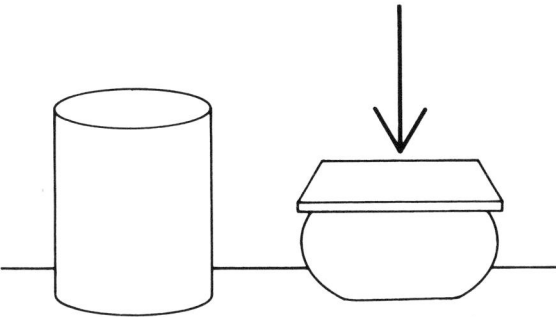

Fig. 1-2. When a force is applied to an object to squeeze it, distortion occurs to accommodate this. Note the decrease in overall height and the "barreling" under compressive load.

only one-half the compressive force and is squeezed 50% less than if one vertebral body alone were present. The strength of each vertebral body has not inherently increased because of its own structure, but the stress on each has been reduced because the supporting area has doubled. This has effectively halved the load to which each segment is subjected. To eliminate the effect of the geometry, we can divide the load by the cross-sectional area and obtain the compressive *stress* or force per unit cross-sectional area. Units of stress are the force per given area: newtons per square meter (N/m² = Pascal [Pa]), pounds per square inch (psi), and so on. Thus, in discussing the intrinsic *strength* of vertebral body material, the mechanical stress present is the relevant quantity.

Suppose we place the same stress on a disc and a vertebral body. What is the resistance of each to being crushed? In other words, how strong is each? Under the same stress, the vertebral body is deformed less than the disc. The bone is stiffer than the disc. Under the same compressive stress

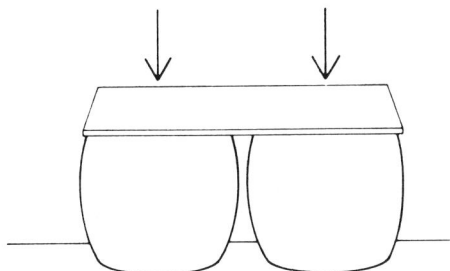

Fig. 1-3. If the same amount of force as shown in Figure 1-2 is distributed over two objects of the same size, each object receives only one-half the load and is therefore deformed one-half as much. Consequently, the compressive *stress* on each is one-half what it would be if only one of the two objects received the entire force.

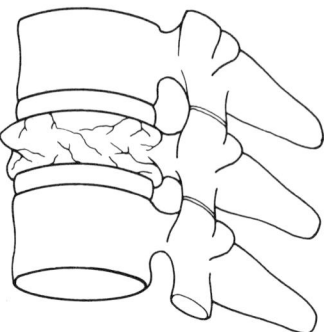

Fig. 1-4. Compression fracture: most common injury to the bony spine.

the difference in the amounts of deformation produced reflects how much more deformable the disc is than the vertebral body. But it can accurately reflect this only if both were of the same height to begin with. Again, to characterize the material rather than the geometry, the total deformation has to be divided by the overall height. The result is the fractional change in length of material and is called a strain, in this case a compressive strain. Strain is usually represented as the percent of the initial length the amount of deformation represents. Ten percent strain means the object has been deformed one-tenth of its initial length.

Since vertebral bodies, discs, or any other objects are compressed only when squeezed, a strain is produced only when a stress arises. A relationship between these two quantities exists. For small strains, the relationship is simple and any given material is characterized by a fixed number that is the ratio of the stress divided by the strain. This number is called the *elastic modulus* (Young's modulus) and is used to relate the susceptibility to deformation of different materials. The higher the elastic modulus, the greater the stress that is needed to produce a given strain (deformation) and the stiffer the material. For example, a vertebral body has a higher elastic modulus than the disc. It is stiffer than the disc; therefore, under the same compressive force, the vertebral body deforms less. Thus, if a person's spine is shortened by two centimeters over the course of a day, it is predominantly a loss in height of the discs since they are more easily deformed than the vertebral bodies.

The most common injury to the spine is a compression fracture of the vertebral body (Fig. 1-4). Since the vertebral body is stiffer than the disc, why does the disc (which deforms more than the bone) not break first? The answer lies in the fact that different materials lose their integrity and break at different strains. The stress at which the material fails is called its *ultimate strength*. *Stiffness* and ultimate strength vary from material to material, but the two are not necessarily directly related. The bone in the vertebral body is approximately 100 times stiffer than the material that makes up the disc. Therefore, a given compressive strain (percent compression) can be achieved 100 times easier (with one-one hundredth the amount of compressive stress) in the disc than in the vertebral body. For this reason, under an applied stress below that required to crush bone, disc compressive strain is 100 times as large as the compressive strain in the vertebral bodies, so that almost all of the actual compressive displacement occurs in the disc (Fig. 1-5). When the discs have deformed to their maximal strain, the stress in both the discs and vertebral bodies continues to rise. When ultimate compressive strength of the bone is reached, the bone begins to crush.

Bone may fracture because of abnormal loads that exceed the strength of normal bone or because of essentially normal loading patterns when the bone is abnormally weak. Often compression fractures occur in porotic bone under the normal compressive loads of everyday life because the mechanical character of the bone has changed. Biologic materials may change in character depending on alterations in normal physiologic conditions (Wolff's law) such as chronic increases or decreases in physical activity or in pathologic conditions (osteomalacia, rickets, idiopathic osteoporosis). These conditions may alter the stiffness of the material (elastic or Young's modulus; ie, the amount of deformation that occurs with any given stress). Such conditions can also reduce the ultimate strength of the vertebral body (ie, the amount of deformation beyond which distortion of the structure is irrecoverable).

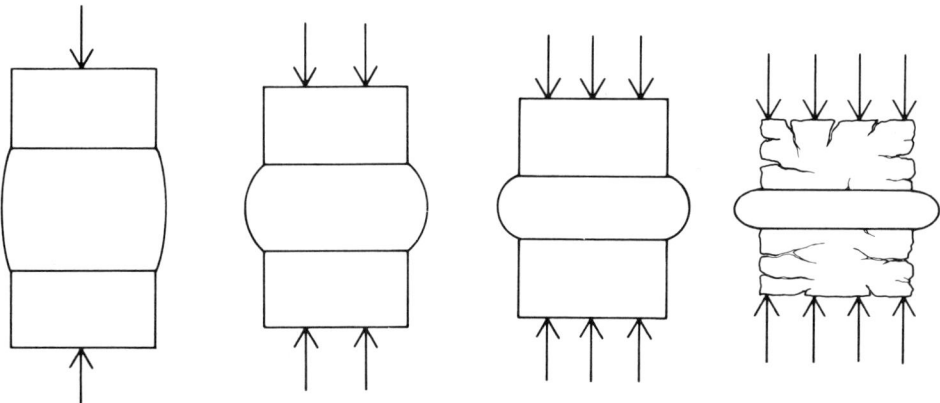

Fig. 1-5. Under low to moderate load, a vertebral body-disc unit primarily deforms in the structure of the disc because it is not as stiff as the structure of bone. At high loads, however, the breaking strength of bone is lower than that of the disc, and hence ultimate deformation is in the vertebral body.

6 PRACTICAL BIOMECHANICS FOR THE ORTHOPEDIC SURGEON

3 MECHANICS OF THE SPINAL COLUMN— NEWTON'S FIRST AND THIRD LAWS

We have considered some properties of the individual components of the spine: the vertebral body and the disc. Let us now consider the basic mechanical properties of the spine.

Compressive stress can be produced only if equal and opposite loads are applied to opposite sides of a vertebral body (Fig. 1-6). This is consistent with practical experience. If one applies an unopposed force to one side of an object, that object moves (is pushed) away. For an object to remain stationary, an equal and opposite force must resist the initial "push." This in essence is *Newton's first law*, which in modified form states that if an object is standing still the sum of the forces acting on it must be zero. Each force applied must be resisted by a force of equal magnitude and opposite direction for the object to remain stationary (Fig. 1-6).

If the vertebrae that support a load are stacked rather than placed side by side, there is no increase in overall compressive strength (Fig. 1-7). The load is transmitted down the column from one vertebra to the next. Each is loaded as if it were subjected to the entire load. As the load is transferred down the spine, the load on each successive unit is slightly increased by the weight of the vertebrae above it. In reality, with descending position, the cross-sectional area of each vertebral unit increases. One could postulate that this may represent an attempt to maintain the same compressive stress at all levels.

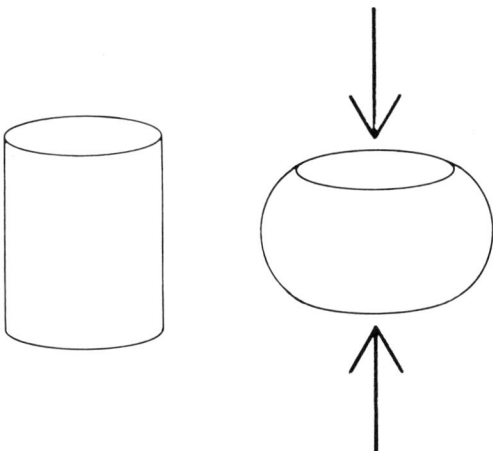

Fig. 1-6. If a load is applied to the top of an object, there will be compressive stress on that object only if an equal amount of load is applied to the opposite side so that the material is squeezed.

3. MECHANICS OF THE SPINAL COLUMN—NEWTON'S FIRST AND THIRD LAWS

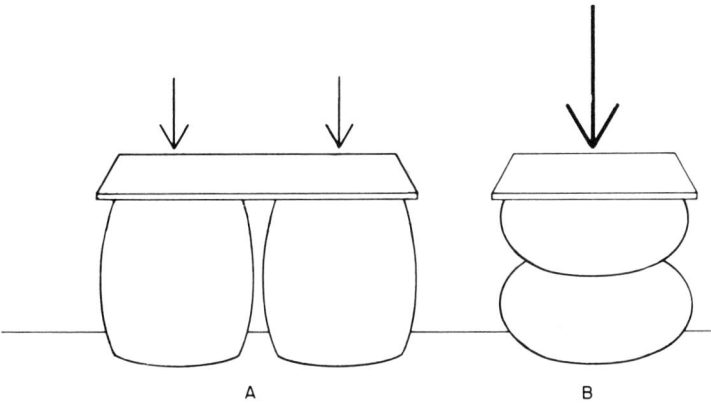

Fig. 1-7. If two objects are placed one on top of the other rather than side by side, and the same amount of force is applied to each configuration, stacking provides a smaller area over which the force can be applied. The cross-sectional area in Fig. **B** is one-half that in Fig. **A;** hence the compressive forces on the top object are twice as much as on each unit in Fig. **A.** In the stack, each unit bears the whole load.

In Figure 1-7B, the load is applied to the top of the upper vertebra represented. From Newton's first law we recognize that an equal and opposite load must occur at the bottom. The latter is associated with the squeezing together of the two vertebrae. Each vertebra is being compressed. The force at the top of the lower vertebra is equal and opposite to the force at the bottom of the upper vertebra. This is *Newton's third law*: for each action there is an equal and opposite reaction.* The action and reaction are the forces exerted by one vertebra on the other and the resulting force upward from the second vertebra to the first. Newton's first and third laws can be used to determine the forces on any bone or musculoskeletal element if sufficient information is available. From the applied forces, the forces on any element and on the remainder of the musculoskeletal system can be determined.

Although the stacking configuration of vertebrae does not change their overall strength in compression, differences in dimensional changes occur. Each vertebra and disc in the column is deformed as a relatively independent unit; if we have 10 vertebrae in the stack, each is deformed the same amount under the same circumstances, since the load on each is nearly identical (Fig. 1-8). For the stack of 10, if the strain of each is 0.1, there is an overall diminution of length equal to the height of one vertebra.

---

*The equal and opposite "reaction" is not limited to force but can include movement. We discuss movements of the spinal segments along with scoliosis, lordosis, and spinal fracture-dislocation later in this chapter.

8 PRACTICAL BIOMECHANICS FOR THE ORTHOPEDIC SURGEON

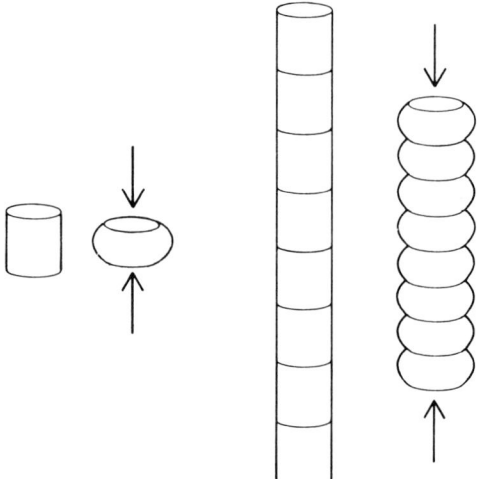

Fig. 1-8. Since each unit in a stacked position has the same amount of load applied to it as every other unit, whatever is produced in a single unit is multiplied by the total number of units involved to give the total deformation of the loaded stack.

In compression, materials will barrel (expand sideways) as their height is compressed. The degree to which the sideways deformation (widening in compression) is related to the change in height (compressed in compression) is *Poisson's ratio*.

4 TENSION

It is common in orthopedics to apply skeletal traction in the treatment of cervical fracture-dislocations. A force is applied to the spine in an attempt to restore the individual segments to their original shape and position. It is a force applied in *tension* (Fig. 1-9).

Each vertebral body, disc, and associated ligamentous structure receives this tension as the traction force is opposed by body weight and initially by muscle spasm acting in the opposite direction. If the supine patient is not pulled toward the head of the bed by the traction, the forces are balanced according to Newton's first law. No motion occurs, and the paravertebral soft tissues are effectively stretched. Under such stretching, discs deform by elongating and narrowing. The paraspinal ligaments are stretched as well. In the area of fracture, the remaining ligaments are not restrained by bony contiguity. They deform and stretch the most. Movement of the vertebral bodies is then guided by these soft tissues, and the fracture is usually reduced.

Any material that is stretched also narrows. The ratio of the strain perpendicular to the deforming force (the narrowing) to the strain parallel to

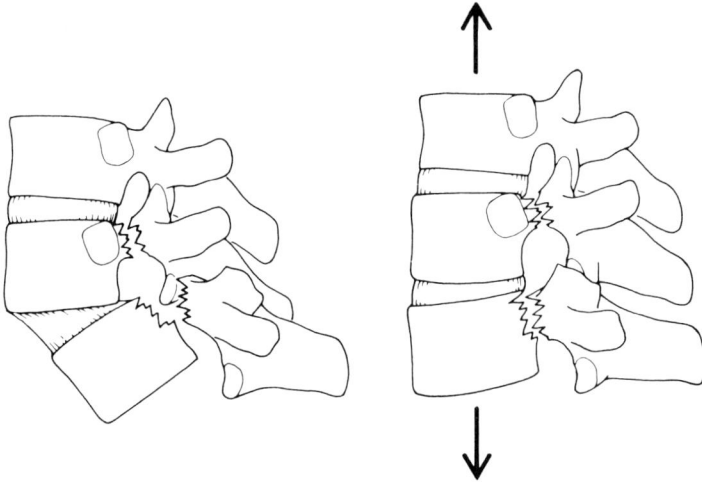

Fig. 1-9. In fresh dislocations of the cervical spine, various ligamentous structures as well as bony units are disrupted. The disrupted structures are not as rigid with regard to an applied tensile force. Under such circumstances, greater deformation occurs in these regions than in adjacent regions. For this reason, not uncommonly the fragments can be realigned and reduced by traction.

the force (the lengthening) is another example of Poisson's ratio. Poisson's ratio exists in both tension and compression. This dimensionless ratio varies from 0 to 0.5, with most biologically important materials having a Poisson's ratio of about 0.3.

All the concepts that apply to compression—stress, strain, and elastic (Young's) modulus—apply to tensile forces as well. *Tensile stress* is obtained by dividing the load by the cross-sectional area, as with compressive stress. Tensile strain is defined exactly like compressive strain. The only difference between tension and compression is the direction in which the load is applied to a given body and the subsequent manner in which that body is deformed. In many materials, particularly metals and ceramics, the amount of deformation (strain) per unit force is the same regardless of whether it is in tension or compression. The ratio of stress to strain is the elastic (Young's) modulus in tension or compression. This means that the elastic modulus is the same for tension and compression in most, but not all, structural materials. For soft tissues the elastic modulus is greater for tension than for compression, while in bone the reverse is true.

5 SHEAR

Traction works by applying a continuous longitudinal force to the patient's muscles and ligaments. This can only be achieved if the patient does not move in the direction of the traction. Newton's first law requires that the

10 PRACTICAL BIOMECHANICS FOR THE ORTHOPEDIC SURGEON

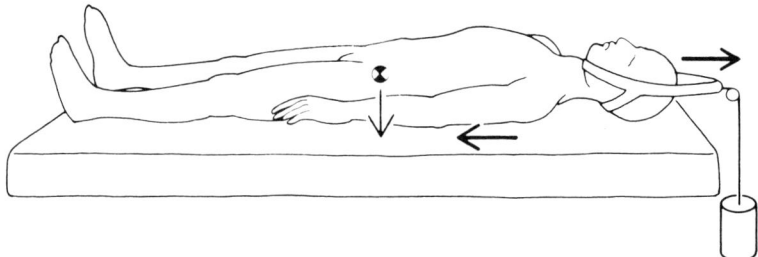

Fig. 1-10. In a supine patient the force applied by cervical traction is resisted by a shear force between the patient's body and the bed. The amount of shear force developed depends on the body weight of the patient. An overweight patient in traction is less likely to slide up in bed than a thin one.

traction force be resisted by an equally strong opposing force. If 50 newtons (about 11 pounds) of cervical traction is applied to a supine patient while in bed, the patient tends to slide toward the head of the bed. Body weight is directed vertically and the traction is horizontal (ie, they are 90° to each other and therefore not equal and opposite). If a series of rollers were between the patient and the bed, the patient would move and no force would develop in the cervical vertebrae. However, if the patient is kept from moving freely, tension develops in his neck. The frictional force between the patient and the bed, a product of the patient's weight and the *coefficient of friction*, μ, resists the body's cephalad movement. This force applied parallel to the body is a *shear* force (Fig. 1-10).

All the concepts of stress (force per unit cross-sectional area) and strain (percent elongation) are equally applicable for shear.

6 VECTORS

Orthopedic surgeons know from experience that the head of the bed must be raised to keep patients who are in strong cervical traction from sliding toward the head of the bed. The reason is that, by providing an upward angle for the body, the body's weight adds to the shear stress holding the patient onto the sheets (Fig. 1-11). By altering the direction in which the shear force is applied, the vertical force of gravity now contributes to the shear force.

The effectiveness of any force depends on the direction in which that force is applied. If the shear force remains in a horizontal direction, consider the effect of applying a horizontal force (Fig. 1-12) versus applying a vertical force (Fig. 1-13). Except for indirectly increasing the resistance between the body and the sheets, a vertically applied force has no direct effect. The

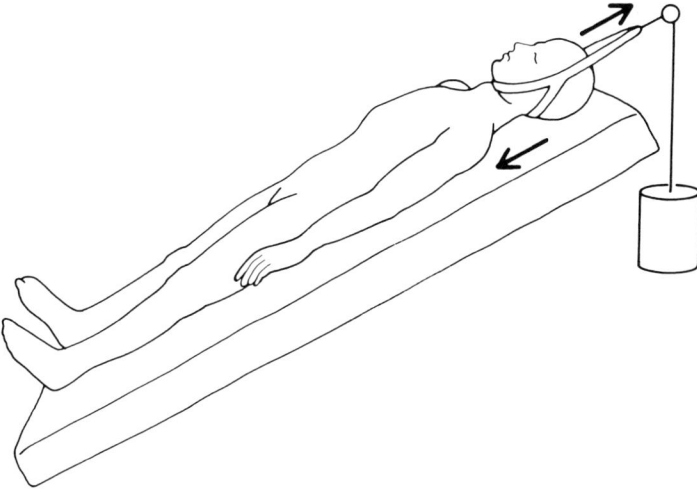

Fig. 1-11. By elevating the head of the bed, a component of body weight is in line with that of the shear forces between the body and the sheets of the bed. It therefore contributes to the restraining forces equal and opposite to the traction forces placed on the head by the halter.

horizontally applied force acting in the same direction as the shear force contributes all of its magnitude to resisting the traction.

Now consider some other force directed at some angle between the vertical and horizontal (Fig. 1-14). Whatever component of that force acts in the horizontal direction creates a shear stress in this case. What *component of a force* acts in a particular direction can be most easily determined graphically, as in the bottom of Figure 1-14. The graphic representation of a force

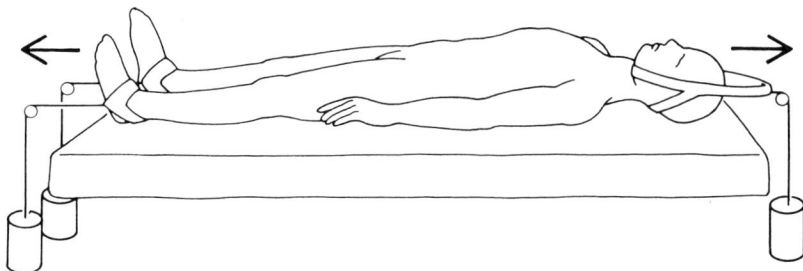

Fig. 1-12. An additional force can be applied to counteract the head halter if traction forces are placed on the torso or legs and aligned parallel with the shear forces of the body against the sheets. In the above case, all of the force of the "leg" traction is added to the shear forces of the body against the sheets.

12 PRACTICAL BIOMECHANICS FOR THE ORTHOPEDIC SURGEON

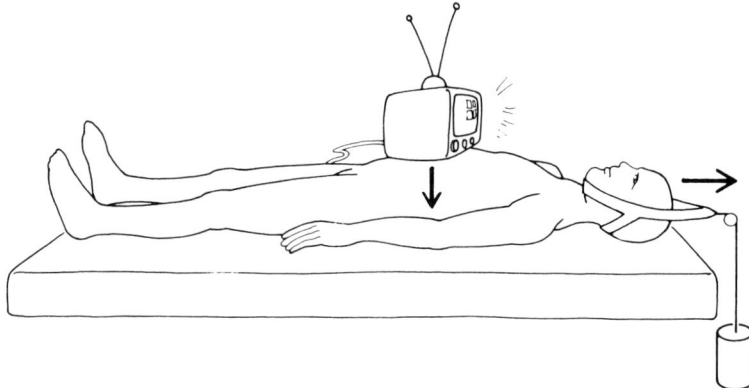

Fig. 1-13. Placing a weight on top of the body, perpendicular to the shear force between the sheets and the body, does not contribute any component of force in a direction to help resist the head halter traction. It does contribute by increasing the friction (shear force) between the bed and the sheets.

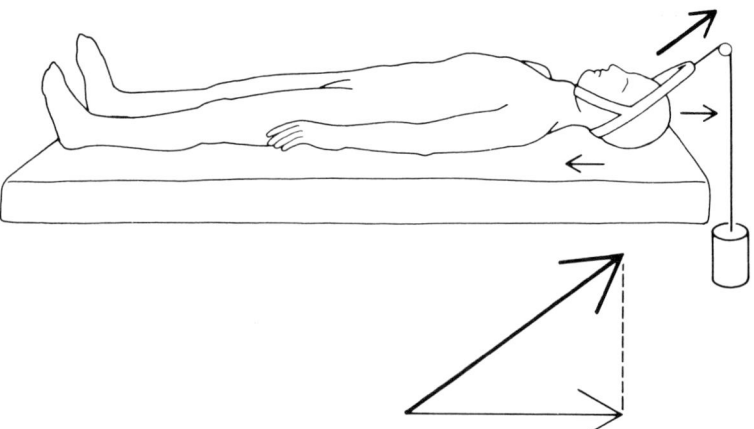

Fig. 1-14. If the head halter traction is applied in a direction that is not parallel to the incline of the body against the sheets, only a component of that force in the direction parallel to the body is resisted by the shear forces of the sheets and body.

or its components is a force *vector*, such as an arrow with its head showing the direction of the force and the length of the arrow proportional to the magnitude of the force.

7 FORCES APPLIED VERSUS STRESSES DEVELOPED

Forward movement of L-5 on S-1 in a patient with spondylolisthesis is produced because of the action of shear forces. The amount of movement is determined by the magnitude of the component of the resultant body weight force in the direction parallel to the vertebral body-disc junction. As the patient leans forward, body weight still acting in a vertical direction produces less compressive force and more shear force at L-5,S-1 (Fig. 1-15).

These increased shear stresses produce movement if an equal and opposite stress is not produced by the disc or ligaments (Newton's first law). Such a degree of resistive stress occurs only when the strain of elongation in these tissues reaches certain limits. Up to that point one body slips forward onto another. When equilibrium is reached, all further movements stop. Hence what appears to be a grade I slip on lying down can become a grade III spondylolisthesis because of the shear forces created during forward leaning.

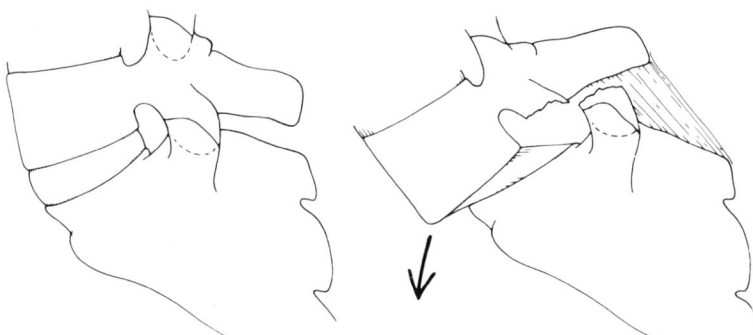

Fig. 1-15. When the body is in the upright position, the vertical force of body weight, when aligned over the vertebral body, is practically perpendicular to the vertebral body disc interface. Under such circumstances, compression between the two units is the major type of stress existing. If, however, the body leans forward, then the vertebral body-disc junction becomes more parallel to the vertical force line of the body weight. In such a circumstance, a substantial component of body weight becomes parallel to the disc-body junction, thus creating shear forces between the two. The force then creates movement unless an equal and opposite force of the same magnitude develops. This occurs only when the ligaments are strained to the appropriate point.

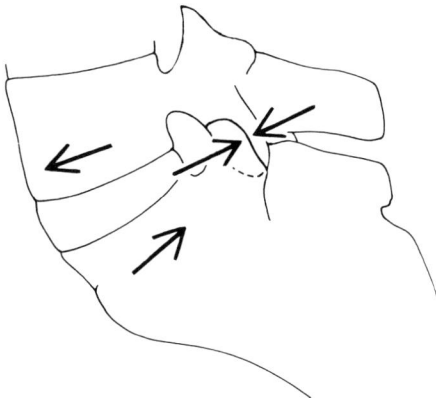

Fig. 1-16. Shear forces that develop as a person leans forward are applied to the posterior facet joints. Since they are aligned perpendicular to the direction of the force, the two sides of the joint are squeezed together. In such a case, they are under compression and movement is prevented because the strength of the bone in compression is sufficiently high.

What happens if there is no spondylolisthesis? What prevents forward movement? Do shear forces still develop in the disc area? As the body bends forward, if the vertebral body and disc alone were considered, the situation would be the same as that noted above for patients with spondylolisthesis. Normally, however, articular facets are connected to the vertebral bodies. As the body bends forward, the same body weight force acting in the direction that would create shear forces on the vertebral body and disc creates compressive forces at the articular facets (Fig. 1-16).

The magnitude of the forces is the same in both cases, except one is counteracted by shear force and the other by compressive force. Since the bone in compression is stiffer than the bone and disc in shear, movement is pre-

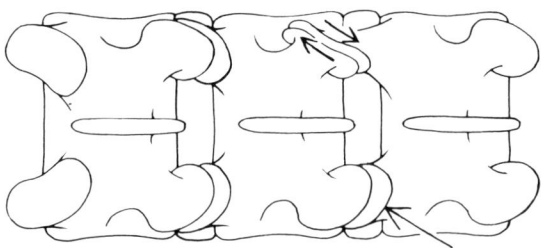

Fig. 1-17. Facets normally convert shear at the disc space into compression across the facet joint. If a facet joint is not perpendicular to the intervertebral shear, then, instead of acting to compress the joint, a component of the shear acts to make the joint slide. Such a malaligned or "asymmetric" joint cannot bear its share of the compressive stress.

vented because for the same force the strain developed in compression is small. No movement occurs. The bone takes all the stress. The strain developed in shear in the disc is therefore small. The articular facets act to resist compression and thus save the disc from significantly deforming shear stresses by not allowing any significant strains to develop. In the case of asymmetric L-5,S-1 facets, when facet alignment is altered, the body weight force as the person bends is no longer directed as a compressive force at these facets but as shear (Fig. 1-17). The disc and ligaments are poorly protected by an asymmetric facet joint.

8 BENDING

Under the same load, then, parts of a vertebra can sustain different types of stresses because of structural geometry. The types of stresses that can be developed in different parts of the disc-vertebral body segment can also differ if the load applied is eccentrically placed. If we return to the L-5,S-1 segment, the examples illustrated above are not quite correct. When the person was standing, the load was said to be located directly over the disc and the vertebral body (pure compression); when the person was leaning forward the load was said to be parallel to the spinal axis and perpendicular to the disc-vertebral body interface (significant shear). Actually, pure compression, tension, and shear are not common in the skeleton. Because of the irregular shapes of the body's supporting structures, applied loads tend to rotate or bend the body, as well as squeeze, stretch, and shear it. Eccentric loading tends to bend the spine (Fig. 1-18).

Bending involves simultaneous tension and compression stresses at different locations within the same construct. When the spine is flexed forward, the interspinous ligaments, ligamentum flavum, and posterior longitudinal ligaments on the posterior convex side are stretched. On the anterior concave side of the spine the disc and vertebral body are compressed. Each is deformed, and the amount of stress produced is in accord with the elastic modulus of the bone, disc, and ligament (as discussed in Section 2).

Bending has therefore stretched the posterior convex side (made it longer) and compressed the anterior concave side (made it shorter). In some plane between the concave side and the convex side there is no stretching or compression (ie, no change in length). At this point, since length is not changed, there is no strain and hence no stress. It is the plane on which the stresses and strains, caused by bending, equal zero. This plane is the neutral plane.

In a simple bar subjected to pure bending (Fig. 1-19), the *neutral axis* is the central plane of the bar—the plane that cuts the bar in half lengthwise. When a column is bent, the fibers on its convex side are stretched and those on its concave side are compressed. From the convex side to the center of the column the fibers are stretched decreasingly; completely unstressed fi-

16 PRACTICAL BIOMECHANICS FOR THE ORTHOPEDIC SURGEON

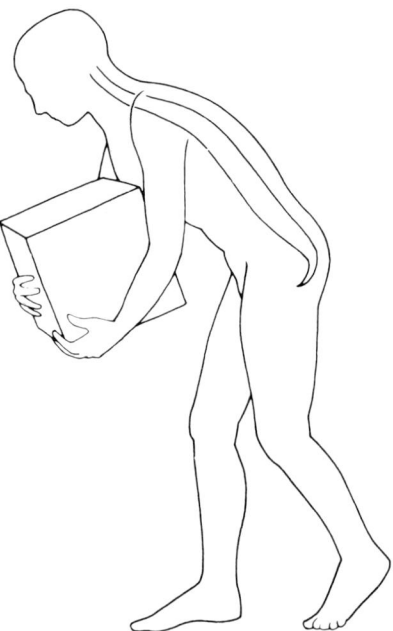

Fig. 1-18. When a person picks up an object, the object is not in line with the structures supporting it. Under such circumstances, the person's trunk is not purely squeezed or stretched, but tends to rotate or bend.

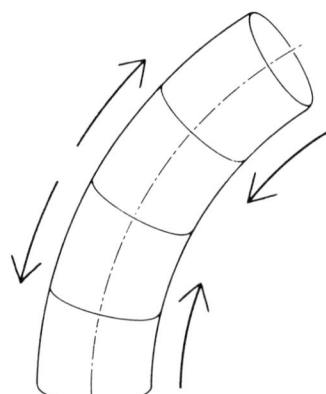

Fig. 1-19. When an object is bent, the area on the convex side is stretched and on the concave side is compressed. At some point the decreasing compression changes over in the material to increased stretching. The line or plane up and down the object connecting all those points is the neutral axis.

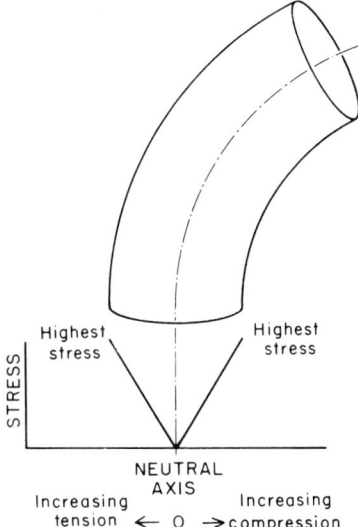

Fig. 1-20. When an object is bent, the further away from the neutral axis the larger the stresses become, since the more the material is either stretched or compressed. On the convex side, the material is stretched and tensile stresses occur; on the concave side, compressive stresses occur.

bers lie in the neutral axis. From the neutral axis outward to the concave side, the fibers are increasingly compressed. From the neutral axis to the concave side, the compressive strain (and stress) increases linearly to a maximum at the surface. The linear increase in strain is just a matter of geometry.

Figure 1-20 shows the stress and strain versus distance from the center line of the bar. The outer surfaces on which the tensile-compressive stresses maximize are the *extreme fibers*. Although the spinal column has no symmetric geometry and is certainly not of homogeneous material, these basic concepts of bending apply. As an example, rupture of the anterior longitudinal ligament and bone at C-5 is usually due to hyperextension in bending. These structures are the outermost fibers in tension and are therefore subjected to the most tension (Fig. 1-21).

It is instructive to think about the stresses present at the lumbosacral junction in bending. When one bends forward, flexing from the upright posture, the amount of compressive strain at the disc may increase. The amount it increases depends on whether the tensile modulus of the ligaments (on the tensile side) is greater than the compressive modulus of the disc (on the compressive side; i.e., whether the ligaments are stiffer than the disc) (Fig. 1-22). At a given deformation (strain), ligaments are stiffer in tension than are discs in compression. The stresses are higher in the ligament than in

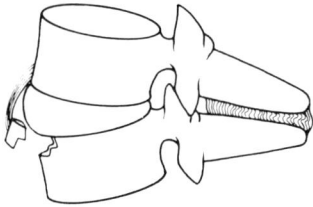

Fig. 1-21. Common fracture illustrating bending occurs after forced hyperextension of the cervical spine. A piece of bone is pulled off by the tensile forces created at outermost fibers of bending. Since the bone is weaker under maximum loads than the ligament, the material that breaks under maximum tension is bone rather than ligament.

the disc because the ligaments are stiffer. The forward bending load of the body is therefore balanced more by the opposing forces in the ligaments than by those in the disc. The posterior ligamentous structures prevent excessive stresses in the anterior aspects of the disc just as the articular facets usually prevent excessive shear stresses in the disc. The implications of this to the production of pain are interesting since, although the disc is aneuronal, the ligaments and facet joint capsules are well innervated, particularly with stretch and pain receptors.

An interesting physiologic implication arises from observing the stress and strains that might be produced in each ligamentous structure when the body bends forward and the ligaments are stretched. If the posterior longitudinal ligament, the ligamentum flavum, and the interspinous ligament all have the same length, cross-sectional area, and modulus of elasticity before bending starts, then the maximum tensile stresses develop in the ligament, having the maximum deformation during bending (ie, the inter-

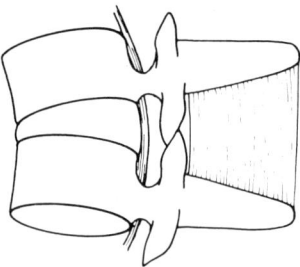

Fig. 1-22. If the amount of material or strength of the material on one side of bending is not the same as on the other side, the stiffest material limits the degree of bending that occurs. If, for example, the ligaments on the posterior elements of the spine are stiffer than the disc, the amount of deformation in forward flexion depends more on the geometry of the posterior elements than on the disc. In such cases, high compressive stresses are prevented in the disc because of the posterior structures.

spinous ligament). This ligament is furthest from the neutral axis and is stretched the most in all attitudes of bending. Since we have assumed that the strength and cross-sectional area of all ligaments are the same, and stress and strain are directly related, then the ligament that is strained the most has the highest stresses. If, however, the modulus of elasticity for each ligament is different, then the tensile stresses that develop in each ligament are determined not only by the amount of stretch present but also by the stiffness of each ligament. Depending on the modulus of each ligament, it is possible that, at one particular point of bending, deformation of each ligament, though different, is such that the stresses in each ligament are the same (ie, the load can be equally shared by all three structures). At other positions of bending, it is possible that a ligament not at maximum deformation may be under the highest stress. For example, if the ligamentum flavum were stiffer than either the posterior longitudinal or interspinous ligaments, even though the ligamentum flavum is stretched to a length somewhere between the two, the tensile stresses might be the highest in this ligament and it would restrict further movement. A similar situation can arise if the cross-sectional areas of the ligaments differ.

An ingenious mechanism by which the body could provide for sharing the stresses between these three structures would be to have different stiffnesses (moduli of elasticity) and cross-sectional areas of each ligament and at the same time allow each ligament to stretch within a different range of its length. Under this situation, the ligaments' strains would not be proportional merely to the distance each is from the center of bending. Such a situation resembles what we know to be the case in real life. The posterior longitudinal, ligamentum flavum, and interspinous ligaments all appear to have different resting lengths. It is interesting to reflect on the changes in stress distribution that might arise if, after laminectomy, any of these tissues is replaced with scar. Scar has a lower stiffness than original ligaments and will be stressed the least. Such a situation probably arises with removal of the ligamentum flavum during laminectomy. Under such circumstances, since some of the stiffness of the overall structure has been removed and replaced by a weaker structure, greater deformations may be permissible. Significant alterations in the stress distribution result. Hence a degree of instability may arise just from laminectomy even without disc removal.

Two other factors protect the disc from increased compressive stresses in forward bending. The first is added protection resulting from active contracture of the paraspinal extensor muscles. With flexion at L-5,S-1, these muscles are active. Since they are eccentrically placed posteriorly, they produce bending in a direction opposite to that produced by body weight (Fig. 1-23). As the body bends forward, the position of these muscles toward the extreme fiber in tension provides them with significant mechanical advantage. If the muscles acted alone, they would create compression in the posterior elements and tension in the anterior elements. Acting in concert with the forward body weight, they neutralize the bending effects of body weight.

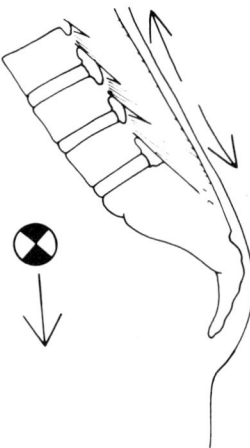

Fig. 1-23. Muscle activity at the most posterior aspect of the spinal structure creates a bending moment opposing that of body weight. Contraction of these muscles tends to compress the posterior elements and stretch the anterior elements and thus nullifies the body weight bending moment. The stresses on the spine created by bending are reduced.

The posterior muscles act to reduce tensile stresses in the posterior ligamentous structures and compressive stresses in the anterior aspect of the spine. This protects the entire structural system from excessive bending stresses. Imagine the increase in stress and strain on these structures when the extensor muscles are weak or when, because of denervation, they cannot act.

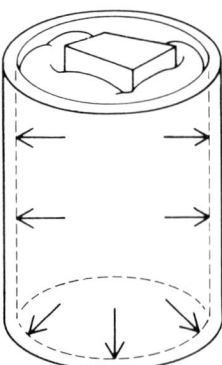

Fig. 1-24. When a load is placed on a constrained bag, it generates an internal, hydrostatic pressure that can support a load.

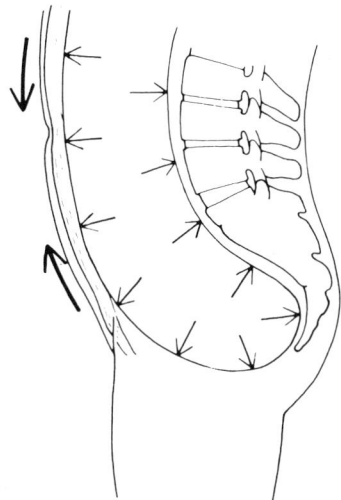

Fig. 1-25. The abdominal cavity may be considered a water-filled balloon. It can support some of the weight of the upper body if lateral deformation of it under load can be prevented by containment. Such constraint is provided by strong contraction of abdominal muscles.

To understand further the mechanics of the spine in bending, consider a pressurized fluid within a rigid walled cylinder. The abdominal cavity can be likened to a balloon placed within a tin can. Compression of the balloon from any direction generates pressure that is equal in all directions. This *hydrostatic pressure* can support loads (Fig. 1-24). The balloon's walls resist its tendency to bulge (or bend) outward. The constrained balloon can support load without further deformation. The abdominal cavity, if its walls do not deform (i.e., when the abdominal muscles firmly contract), can support some weight of the body above (Fig. 1-25).

In a similar manner, the anulus fibrosis acts with the nucleus pulposus to support the compressive load across the intervertebral interspace (Fig. 1-26). When the abdominal cavity acts in the manner shown, the load that the spine must bear is decreased. When a person bends or lifts a heavy object, the abdominal muscles contract as in a valsalva maneuver. Imagine the effects on the load the spine must directly bear with abdominal muscle laxity after stretching from obesity, multiple abdominal procedures, or pregnancies. The effect of the corset as a spinal support is obvious. This also explains the compressive strains that develop in the nucleus pulposus if the anulus fibrosis is allowed to bulge or is weakened by a rent or a tear in its structure.

22 PRACTICAL BIOMECHANICS FOR THE ORTHOPEDIC SURGEON

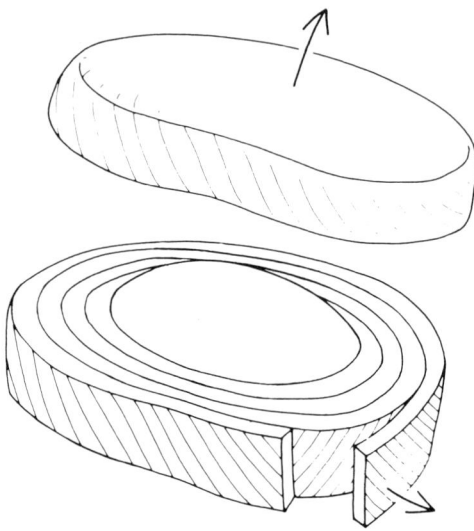

Fig. 1-26. The nucleus pulposus and the anulus fibrosus operate much like a balloon contained within a can. Hydrostatic pressures are created in the nucleus.

9 STRESS CONCENTRATION-EFFECT OF DISC DEGENERATION AND SPINAL FUSION

An important role of the discs is to distribute the strains in the vertebral bodies and spinal column evenly. The disc's modulus of elasticity is much lower than that of a vertebral body; therefore, if a given amount of bending of the back is achieved, it is primarily through the strain produced at the discs. Each disc acts as an articulation and minimizes the bending stresses of the bony portions of the spinal column by allowing the spine to flex at much lower stresses. If a disc degenerates, the space between vertebral bodies narrows and little motion within the segment occurs. What motion does occur is at the posterior elements. Instability results until the deformation creates sufficient stresses to resist further motion. Loss of disc function means greater deformation and greater bending stresses at adjacent discs and vertebrae during a given flexion. A *stress concentration* is thereby produced because of disc degeneration.

Stress concentration occurs when the rigidity of a section of the vertebral column is increased by disc degeneration, fusion, or internal fixation. The greater stresses increase the chances of spondylosis (degeneration) or spondylolysis (instability) in the interspaces at the level of the degenerated disc or above or below a fusion (Fig. 1-27).

The greater the number of rigid segments, the more the remaining segments must deform to achieve the same motion. The more segments fused, the higher the stresses in the adjacent unfused segments. If a plate or rod

9. STRESS CONCENTRATION-EFFECT OF DISC DEGENERATION AND SPINAL FUSION

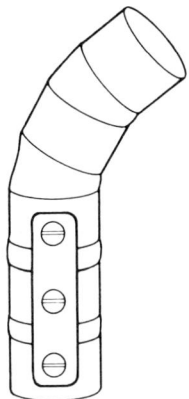

Fig. 1-27. If the motion of several vertebral units is restricted by internal fixation or fusion, the same degree of deformation that might occur over five or six segments can occur over a few segments. Greater deformation results at each remaining level but, most particularly, the greatest deformation is at the segments just adjacent to the stiffened area, and the stresses in these areas are increased.

is affixed to the spine, the metal is more rigid than the bone. Stresses increase in the surrounding interspaces. Stainless steel or chrome-cobalt alloys, cross section for cross section, are 12 times stiffer than bone. If the internally fixed spine is strained to achieve the same kind of displacement, severe bending stresses in the unreinforced part of the spine exist. Making one section rigid considerably increases stresses in the remaining sections. Traumatic spondylolisthesis just below a Harrington rod fusion can be explained on biomechanical grounds (Fig. 1-28).

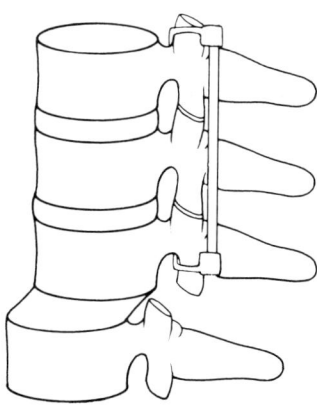

Fig. 1-28. Internal fixation with Harrington rods and spinal fusion for scoliosis produces a stress concentration at the level immediately adjacent to it which may lead to a spondylolisthesis.

10 TORSION

We have discussed compression, tension, shear, and bending as they affect the spinal column. One other type of stress has clinical significance for the practicing orthopedic surgeon—*torsion*. Torsion stresses in the spine are produced by twisting forces that tend to rotate the spine about its long axis. When such forces are applied at any vertebral level, that level tends to rotate about the level below (Fig. 1-29).

Motion occurs at the discs as they deform more for given stress (because of lower Young's modulus) than either vertebral body above or below. As torsional forces are applied, the relative motion of one vertebra about the other creates tensile and shear stresses in the annulus fibrosus. As in bending, the greatest stresses occur in the areas furthest from the center of rotation (the neutral axis for torsion). As they are the most distant from the center of rotation and thus have the longest lever arms, the facet joints again act to spare the discs from excessive stresses and strains. Unlike bending, rotation produces asymmetric stresses at the joints. With rotation, one facet joint closes up and the other opens. Thus compressive and shear stresses are concentrated at one joint whereas tensile stresses are concentrated in the capsular and ligamentous structures of the contralateral joint. This is in contrast to similar stresses created in both facet joints by bending.

Frequently, rotatory injuries of the spine are associated with facet fractures that break in the act of blocking excessive rotation. The common torsional injury of the spine is fracture dislocation of T-12,L-1, the thoracolumbar junction (Fig. 1-30). Above this level the thoracic spine is relatively stiff. The facets in this region are relatively horizontal and offer little resistance to torsion except through tensile stresses in the capsule and ligaments. Additional resistance to torsional stresses are produced by the rib cage and the costovertebral ligaments, which are far away from the neutral axis of torsional rotation. Below the thoracolumbar junction, the lumbar region gains increased resistance to deformation by progressive vertical orientation of the facet joints. The T-12,L-1 region, being a transitional area, has neither the thoracic supplementary protective elements nor the lumbar protective bony geometry; hence torsional deformation and stress are concentrated at the thoracolumbar junction.

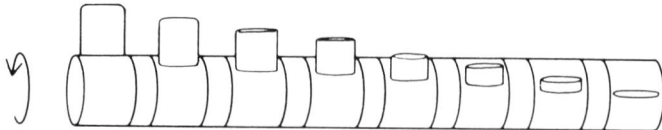

Fig. 1-29. Torsion can be defined as a force applied tending to rotate, or twist, a bar about its long axis.

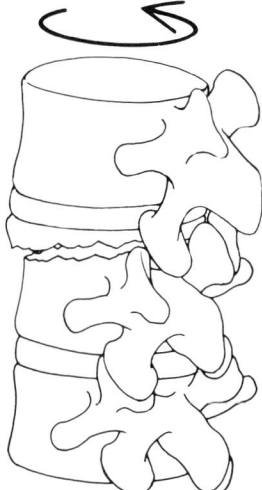

Fig. 1-30. Torsion applied to the spine is transmitted from vertebra to vertebra primarily through the facet joints, which are compressed on one side and pulled apart on the opposite side. If the torsional force is sufficiently high, bony breakage occurs on the compression side and ligaments rupture on the tensile side.

11 LUMBOSACRAL FLEXION EXERCISES AND SPINAL BRACING

From a mechanical point of view, repeated flexion of the lumbosacral spine, such as in the Williams Exercise Program, is contraindicated for patients who have spondylosis or spondylolisthesis at L-5,S-1. Such exercises tend to increase the bending stress at this interspace and aggravate whatever structural problems exist. More appropriate treatment would be abdominal strengthening exercises (Fig. 1-31) and/or abdominal support with a corset or brace.*

Adequate bracing of the lumbosacral spine must limit both flexion-extension and lateral tilt if it is to immobilize the L-5,S-1 interspace completely. Braces that do not have trochanteric horns cannot significantly limit lateral tilt (Fig. 1-32). This probably explains the unreliable results that chairback braces provide in the treatment of L-5,S-1 disease when lateral tilt causes symptoms. The Norton-Brown brace, which provides trochanteric horns, is more effective for the treatment of these conditions. When flexible pelvic obliquity exists secondary to a leg length discrepancy, the pelvis must

---

* The exercises should be done so as not to flex the lumbosacral junction.

26 PRACTICAL BIOMECHANICS FOR THE ORTHOPEDIC SURGEON

Fig. 1-31. Abdominal strengthening exercises should be done in the recumbent position with the lumbosacral spine splinted by the floor or mat. In this position the forces effected by body weight do not act to counter abdominal contraction or create bending moments in the lumbosacral spine.

be balanced with a heel lift or other methods. Pelvic obliquity creates a lateral bending component that puts increased compressive stress across one facet and increased tensile stress across the capsule and ligaments on the opposite side. This requires an increased force from the paraspinal muscles on the tension side sufficient to neutralize this bending stress. This can be a source of pain.

Fixed pelvic obliquities are most commonly associated with scoliosis. The curvature often balances the spine above the pelvis. In such circumstances

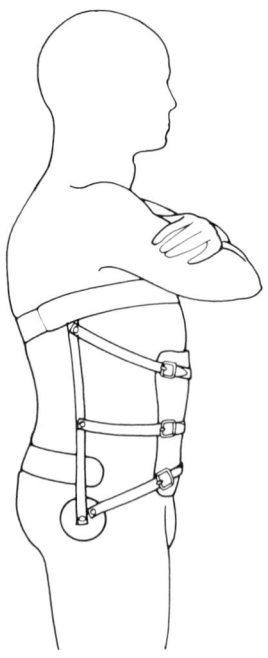

Fig. 1-32. The trochanteric pad of a lumbar brace with rigid sides prevents lateral bending.

a lift under the shoe may only aggravate the symptoms. A heel lift does not alter a fixed pelvic obliquity. If pelvic obliquity is fixed and a fusion of the scoliosis is performed, the lift causes the trunk to be unbalanced and creates a bending moment. High bending stress concentrations will occur below the fusion, and subsequent L-5,S-1 degeneration is likely. In such cases strong consideration should be given to inclusion of the lumbosacral junction in the fusion both to maintain the spine centrally over the sacrum and to prevent late degenerative changes.

12 SPONDYLOLISTHESIS AS A FATIGUE FRACTURE

We have described failures of the spine under various conditions—when the load is too high for the normal structural material to withstand it or when the structure's material properties have weakened, enabling normal forces to cause failure. Can failure occur under normal forces on normal-strength structures?

There is a form of spondylolisthesis, common in adolescents who perform in contact sports or gymnastics, that is a slowly developing fracture of the pars interarticularis. The mechanism of injury is likely impact loading while the L-5,S-1 interspace is repetitively flexed and extended (Fig. 1-33). This

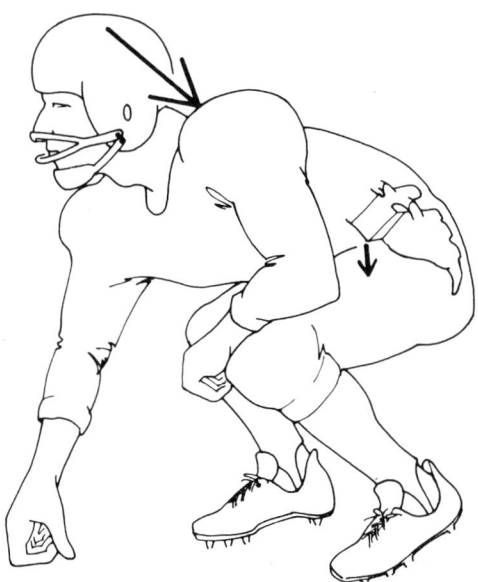

Fig. 1-33. Movement of a player from a crouched position would elicit a lordotic type of bending in the lumbar spine. Additionally, if the subject is struck in the shoulder or chest region, a significant extension bending moment is created. Repetitive reversals in flexion-extension under such circumstances may cause failure in the area of the pars.

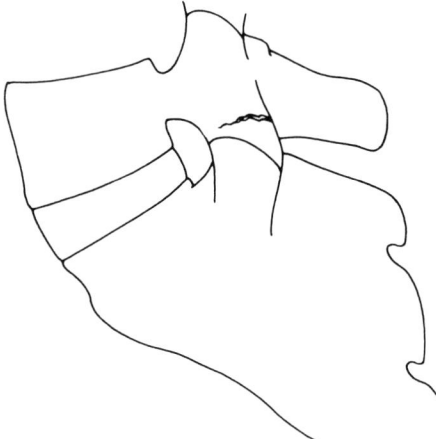

Fig. 1-34. In the pars, a fatigue fracture can occur from tensile stresses attempting to separate the bone. Once a crack is started, each flexion and extension further opens it, leading to eventual failure.

has to be done on a repetitive basis because the pars interarticularis, unless it has some congenital anomalies, can probably withstand the stress induced by a single normal impact. It is thought that with repetitive, frequent loads the bending stresses eventually produce a small crack. This initially occurs on the tensile (posterior) side of the pars and propagates slowly with each repetitive insult until it works its way completely across the bone (Fig. 1-34). It is like breaking a paper clip by repeatedly bending it back and forth. This is known as fatiguing and the subsequent result is a *fatigue fracture*, a ubiquitous engineering problem. More is said about fatigue failure in Chapters 2 and 3.

Such fatigue fractures probably appear at the pars because the tensile stresses in this area of the bone are the greatest. When spondylolisthesis results from impulse loading causing a fatigue fracture, avoidance of these repetitive bending stresses may allow the fracture to heal (if it is still in the incomplete stage). However, once the pars has ruptured, it is relatively unstable, and conservative measures that bring about healing in a stable fracture may not work.

13 RESISTANCE OF THE SPINE TO BENDING

The spine can be described as a long, flexible rod. In bending, a rod is deformed by vertical *buckling* forces. The vertical force that causes buckling is the *critical load*.

13. RESISTANCE OF THE SPINE TO BENDING   29

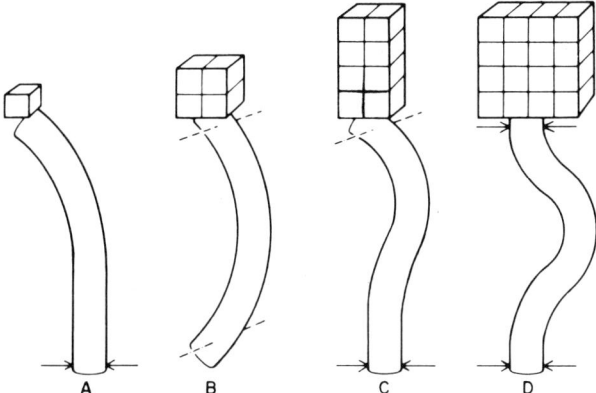

Fig. 1-35. The amount of load a column supports without buckling depends on how the column is restrained.

In Figure 1-35, various constraints have been applied to a rod that is vertically loaded and attached to a base. If no restraint to lateral bending is applied at the top, a minimal load is necessary for buckling and a C-curve results (Fig. 1-35A). If lateral motion at the ends of the rod are restrained, but the top and bottom are allowed to rotate, the force necessary to buckle the rod is increased by a factor of 4 (Fig. 1-35B). Restraints that prevent lateral deviation, but allow rotation at the top, further double the critical load (Fig. 1-35C). This corresponds to the human situation where lateral deviation is restrained but the upper end of the spine can bend freely while the lower end is fixed through the sacrum to the pelvis. Compared with the unsupported model shown in Figure 1-35A, this configuration requires eight times the load to buckle. The critical load can be maximized by creating a situation in which both lateral deviation and bending at the top and bottom are prevented (Fig. 1-35D).

The human spine is naturally bent in the sagittal or lateral plane (postural kyphosis and lordosis) (Fig. 1-36). When scoliosis (bending in the frontal plane) occurs, spinal rotation becomes necessary. This is true because one cannot bend a flexible rod in two planes perpendicular to each other without rotating the rod. The most flexible area of the spine in rotation has been shown to be around T-7. The lumbar facets, because of their more vertical orientation, tend to limit rotation more than the thoracic facets. Hence, in scoliosis, the greatest rotatory deformities occur in the thoracic region.

To a great extent, the flexibility of the spine results from intervertebral discs that allow bending, rotation, and a limited amount of tilt and sideways slip of the vertebrae. Movement beyond these limits requires interruption of the disc and its associated ligaments. The discs contribute much of the flexibility of the spinal column as well as a major restraint to spinal motion.

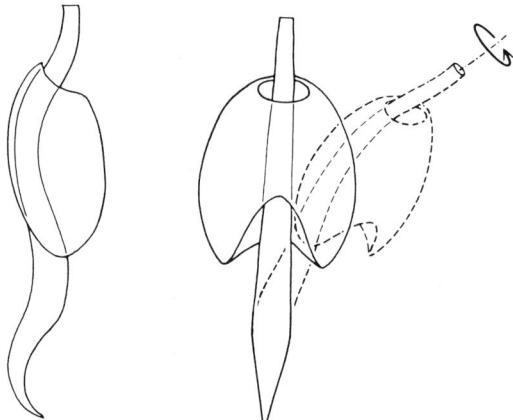

Fig. 1-36. The normal spine has a thoracic kyphosis and a lumbar lordosis. The superimposition of a lateral curvature (scoliosis) on this configuration results in an obligatory axial rotation of the spinal column as well.

The resistance of the spinal column, vertebrae, discs, and ligaments to buckling is small. It requires only about 20 newtons (4.5 pounds) to buckle a thoracolumbar spine that has been dissected from an adult, human cadaver.

However, the critical load that buckles the adult human cadaver spine with ribs and sacrum still attached has been calculated to be about 350

Fig. 1-37. In a patient with scoliosis, muscles contracting on the concave side in normal daily activities can cause accentuation of the curve. Muscles on the opposite side may tend to neutralize the curve and attempt to unbend it, but are usually at a mechanical disadvantage.

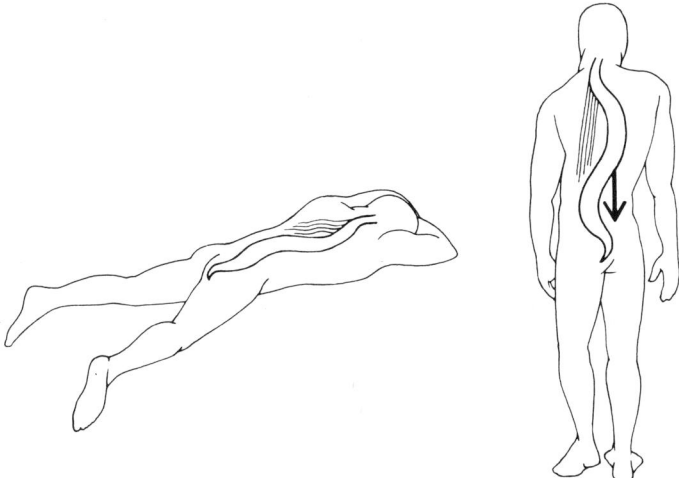

Fig. 1-38. Forces to correct scoliosis would be most effective when resistance is minimal. This would occur in the recumbent position, because, when the subject is upright, body weight produces bending moments that tend to aggravate the scoliosis.

newtons (about 80 pounds) of force. This is also the average combined weight of the upper torso, head, and arms of a normal adult. Stability of the spine against buckling in vivo must obviously involve factors additional to the intrinsic anatomic structures. These extrinsic stabilizing factors are the forces exerted by the trunk muscles. Generally, in a normal spine, symmetric contraction of these muscles prevents buckling. However, when the spine is rotated and laterally bent, the muscles located lateral to the midline can act as further deforming forces (Fig. 1-37).

Once an imbalance occurs, the misaligned muscular forces, or absence of muscular forces as in paralytic situations, added to the forces produced by the body mass cause the curve to progress further. If this occurs in a growing child, accommodation to the abnormal stress causes uneven epiphyseal growth, vertebral wedging, and permanent bone curvature. Even after growth has ceased, the abnormal muscular forces acting on an adult scoliotic spine, through the process of soft tissue remodeling, can cause further progression of curves in a slow but significant fashion. The mechanical factors that induce scoliotic deformities would therefore be expected to be most effective in the upright posture (Fig. 1-38).

14 MECHANICS OF STRAIGHTENING A CURVED SPINE

The principle of straightening a curved spine is to unbend it. Consider the forces that are needed to unbend a simple C-curve. Like the C-curve, the spine can be straightened by placing it in traction or by pushing on it lat-

32 PRACTICAL BIOMECHANICS FOR THE ORTHOPEDIC SURGEON

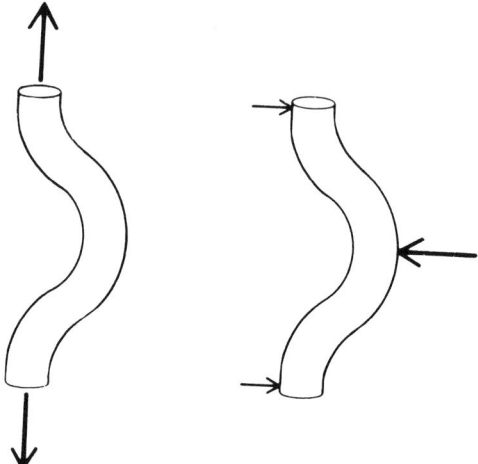

Fig. 1-39. To be effective, both traction and pushing on a curved spine to straighten it require equal and opposite forces.

erally on the apex of the curve. The corrective force must be opposed by equal and opposite forces (Fig. 1-39).

Thus to straighten the spine you must push or pull against some resistance. Any unopposed force, either a pull at one end of the curve or a push at the apex of the curve, just accelerates the spine but does not straighten it. Remember Newton's first law! Equal and opposite forces must be applied so that the ends of the curve are free to move away from each other as straightening progresses (Fig. 1-40); otherwise, only the pattern of curvature may be changed rather than straightening the spine.

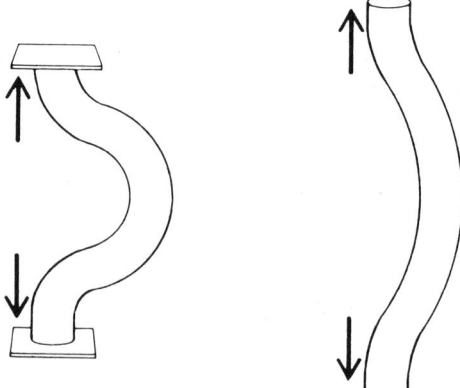

Fig. 1-40. If forces are applied to a curve but movement is prevented, all attempts at correction are futile. Movement must be allowed.

14. MECHANICS OF STRAIGHTENING A CURVED SPINE

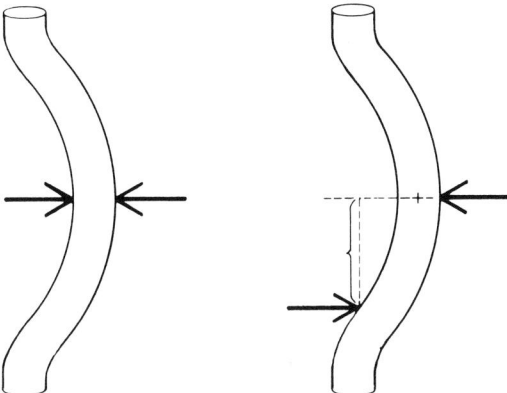

Fig. 1-41. Equal and opposite forces applied at the same level on a curve have no rotatory (straightening) effect. For straightening (unbending) to occur, the equal and opposite forces have to be applied a distance apart. The magnitude of the (un)bending moment applied depends on the distance of these forces from the center of curvature.

To straighten a scoliotic curve, forces must act at some distance away from the apex of the curve (Fig. 1-41). The perpendicular distance from the line of application of the force to the apex of the curve is defined as the lever arm or *moment arm* of the force. The longer the lever arm, the greater the counter-bending produced. Counter-bending effectiveness is determined by multiplying the force by the length of the lever arm. The product of this multiplication is defined as the moment of the force or the *bending moment*.

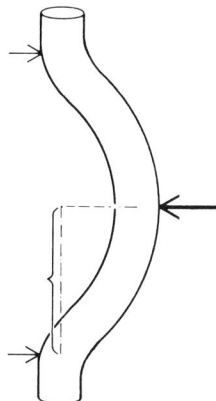

Fig. 1-42. The force on the right-hand side of the curve is equal and opposite to the two forces on the left-hand side. No lateral motion occurs. Since the forces are equal and opposite but are not in the same line (i.e., spaced a distance apart) the two tend to rotate the top part of the curve clockwise, and the bottom two tend to rotate the bottom part of the curve counterclockwise, thereby straightening the curve.

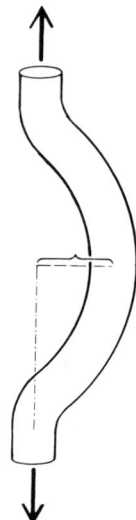

Fig. 1-43. Another configuration producing the same result as in Figure 1-42. Here, both forces are off center; therefore, the top force produces a clockwise bending moment on the top part of the curve, and the bottom force produces counterclockwise bending on the bottom part of the curve. Since the two forces are equal and opposite, the structure does not move up or down but merely straightens.

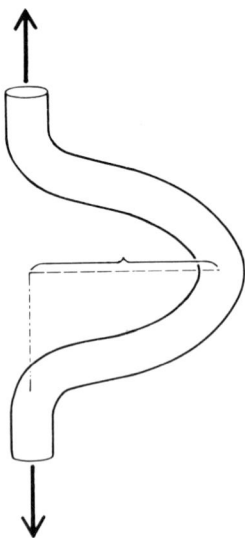

Fig. 1-44. In comparison to Figure 1-43, the same forces applied on a severe curve have a greater tendency to unbend the curve, since the moment arm of each force from the center of rotation is large.

14. MECHANICS OF STRAIGHTENING A CURVED SPINE

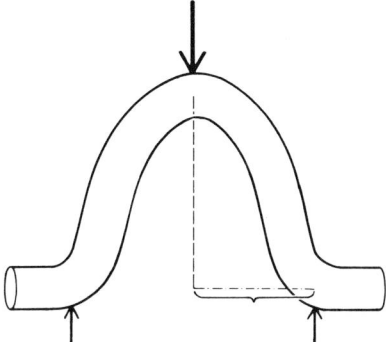

Fig. 1-45. If forces of similar magnitudes as in Figure 1-44 are applied in opposite directions, there is less of a tendency for the curve to be straightened since, although the magnitude of the forces is about the same, the moment arm of each from the center is less.

Bending moments are created by (1) pushing in separated places along the spine (Fig. 1-42) or (2) pulling on the ends of the spine (Fig. 1-43). Consider these two methods in relatively mild curves. Horizontally applied forces create significant bending moments because their lever arms are relatively large (Fig. 1-42). To be effective, traction forces must be very large because the lever arm they work through is relatively small (Fig. 1-43).

In severe curves, traction forces are more efficient because they have significantly greater lever arms (Fig. 1-44). Horizontally applied forces are at a mechanical disadvantage (Fig. 1-45), or at least a lesser advantage.

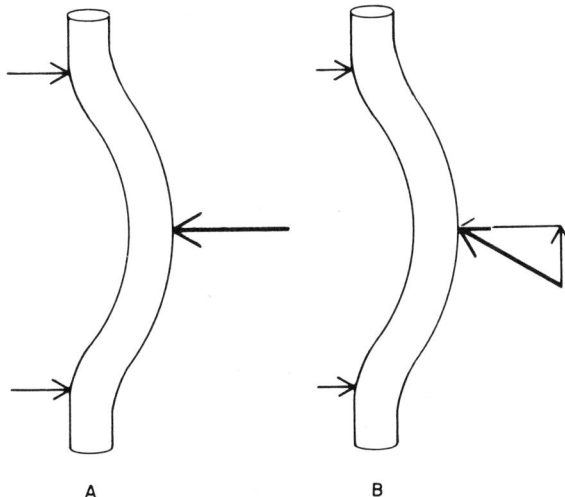

Fig. 1-46. A force applied perpendicular to the curve (as in Fig. **A**) is most effective in straightening the curve. If a force is applied at an angle (as in Fig. **B**), only the lateral component of the force contributes to the bending moment.

36 PRACTICAL BIOMECHANICS FOR THE ORTHOPEDIC SURGEON

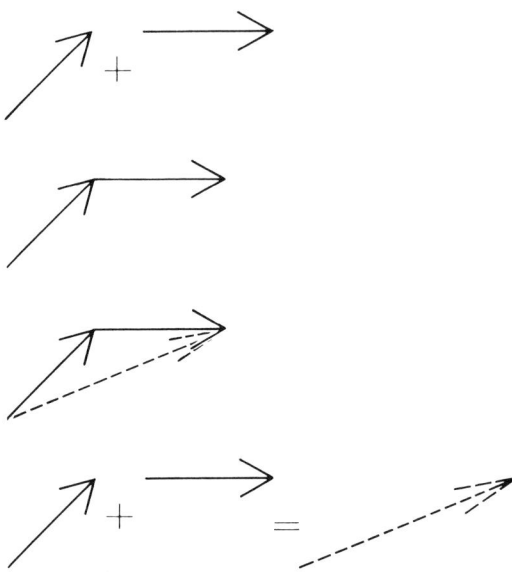

Fig. 1-47. When several forces acting on the same body are directed at different angles, the magnitude and direction of the total resulting force acting on that body may be obtained by placing each force head to tail against the other ones, maintaining its direction and representing its magnitude by the length of the line. The resultant force is the line connecting the tail of the first force to the head of the last force.

We have discussed the magnitude of relevance of applied forces, their point of application, and the lever arms through which they function in creating a bending moment. The direction in which the force is applied is also a determining factor in creating a bending moment. Compare a horizontally applied counterforce to one of similar magnitude and point of application that is skewed from the horizontal (Fig. 1-46). Only the horizontal component of the force helps to straighten the curve. The vertical component of the pushing force has no beneficial effect in this case.

The component of a force that acts in a particular direction can be most easily determined graphically.* When lateral and traction forces combine to correct a curve, both vertical and horizontal forces act at the ends of the spine. One can add the moments of the components that make up the *resultant force* in the appropriate directions. The resultant bending moment

---

* Our intuitive isolation of the spine to study the effect of the forces acting on it is valid from an engineering standpoint as long as we consider all the forces and counterforces at work on this isolated structure. The creation of such an isolated static equilibrium is referred to as *free body analysis*.

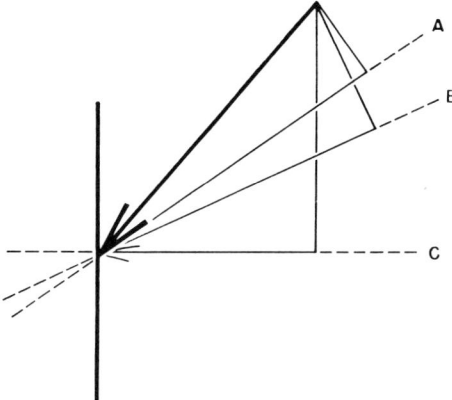

Fig. 1-48. (A,B,C) A given force has a component in any particular direction by drawing a right triangle. The base of that triangle then gives the magnitude and direction of that component.

acting on the spine is the sum of the bending moments contributed by each component of the forces acting in a meaningful fashion on the spine (Fig. 1-47). The *components of a force* that act in a particular direction can be determined graphically (Fig. 1-48).

15 TRACTION, CASTS, AND BRACES

Traction, casts, and braces either pull or push the spine, correcting scoliosis by creating bending moments. The spine can be pulled by skeletal halo-femoral or halo-pelvic traction (Fig. 1-49). Theoretically, traction is effective in severe curves if they are flexible. However, many curves are stiff and inflexible, and the substantial traction forces necessary to overcome them cannot be generated because of the holding power limitations of the pins in the bone or the amount of traction the ligaments and discs in the flexible noncurved areas can resist. As a result, combinations of soft tissue releases and skeletal traction are often used to severe curves.

A scoliotic spine can also be straightened by applying a bending moment to the curve with a turnbuckle jacket (Fig. 1-50). In this technique, a body cast is applied with both ends of the spine fixed in the plaster. Hinges are plastered directly over the apex of the curve. The cast is then cut so that the top and bottom can be spread using a turnbuckle. This essentially applies counter-bending to the spine with relatively large moment arms. This technique is biomechanically very effective. It is only limited by tissue tolerance to the applied bending moments.

A more recent modification of this technique, the localizer jacket, is an attempt to improve the turnbuckle jacket. A body cast is applied while the

38 PRACTICAL BIOMECHANICS FOR THE ORTHOPEDIC SURGEON

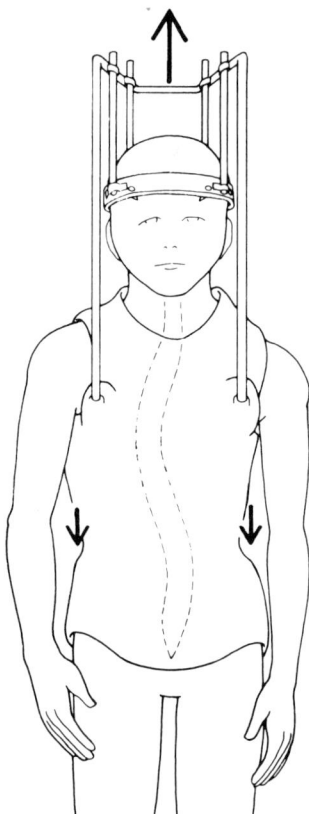

Fig. 1-49. Halo-femoral or halo-pelvic traction by fixation to the skull and bony pelvis or femurs. Traction then produces equal and opposite tensile forces that tend to unbend the spine.

spine is in traction, and lateral pressure is added to the curve through the pusher pads (Fig. 1-51). This combination is mechanically advantageous because it corrects both the severe curve (traction) and the curve as it becomes milder (lateral pressure). As correction is obtained, new casts can be applied serially.

This method has been further modified for more practical use by the Milwaukee brace. This device combines traction and pushing forces such that they can be adjusted as the curve corrects while the patient grows (Fig. 1-52). The traction from the Milwaukee brace is mostly achieved by patients' attempts to stretch out of the brace with their own muscular activity. The brace prevents lateral deviation and bending and establishes constraints that require the greatest critical load to buckle the spine (see Fig. 1-35D).

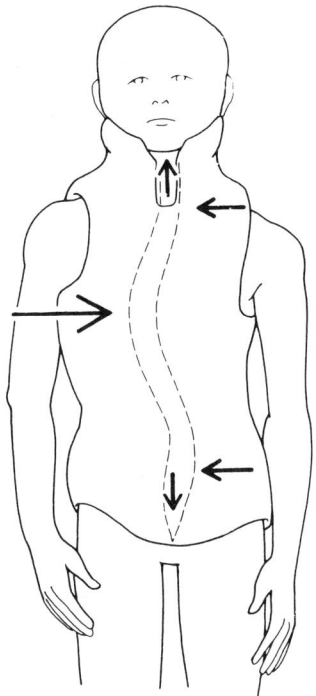

Fig. 1-50. A turnbuckle jacket produces equal and opposite forces that, because they are placed off center and have significant moment arms, can unbend a scoliotic curve.

This mechanical situation acts to limit the progression of scoliosis, as gravity and other deforming forces must be maximal to buckle the spine. Lying down obviates these deforming forces and allows the brace to be most efficient. Thus the most effective situation for correction exists when the patient wears the Milwaukee brace in bed.

If the extrinsic stabilizing forces from muscular contractions are absent, minimal forces can buckle the spine. The brace cannot prevent this. This is because the corrective forces that are required cannot be tolerated. At the other extreme, the muscle strength in patients with scoliosis secondary to spasticity is usually too great for a Milwaukee brace to overcome. If the stabilizing forces are not functioning, progression of the curve within the brace is likely. Using a Milwaukee brace, then, would not be effective in treating patients with paralytic curves or who cannot actively stretch out of the brace (and/or in situations in which the internal forces are high). Such curves are more responsive to casting or jackets, which act by external con-

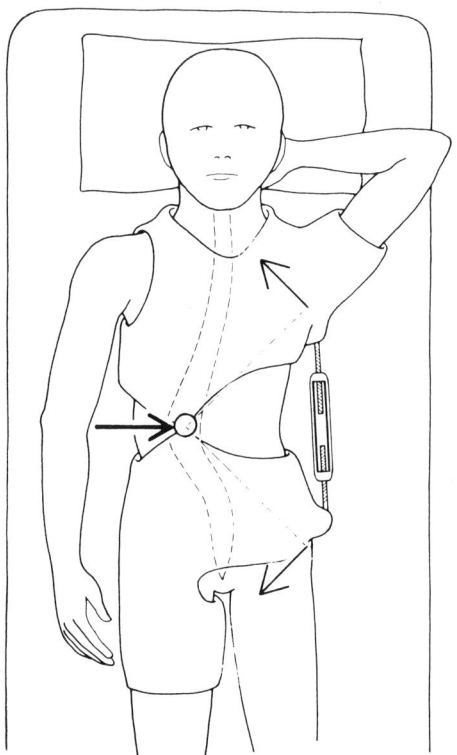

Fig. 1-51. A localizer jacket combines tensile traction through the head and pelvis with lateral forces to create bending moments from different directions.

tainment as a bucket around the spine. This creates a condition of maximum constraint to lateral bending and free rotation (Fig. 1-53).

The Milwaukee brace is most effective if the curve is small and the rib pad is arranged almost perpendicular to the spine. The force of the thoracic pad is transferred to the vertebral spine through the ribs as they attach to the vertebral body and transverse processes. Apically placed thoracic pads perpendicular to the long axis of the spine and opposing forces positioned laterally at the pelvis and neck act through long lever arms and create significant moments with relatively small forces.

If the curve is large, the force created by the pad and neck and pelvic supports tends to become more parallel to the axis of the curve and less corrective. Furthermore, such patients usually have a severe rib deformity. The pad would have to be applied in a more vertical direction because horizontal application would only push the ribs inward (Fig. 1-54).

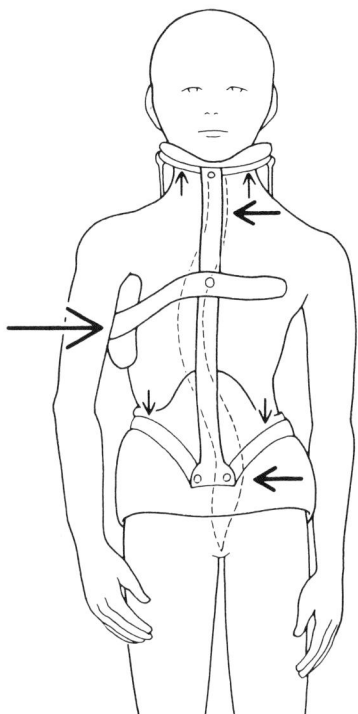

Fig. 1-52. A Milwaukee brace has the potential to produce the same type of forces in the same manner as does a localizer jacket. Because the forces are applied at specific points, high pressure may limit the amount of force that can be generated. By the body reacting by "withdrawal" from these contact forces, a similar effect to unbend the spine can occur from muscle activity.

Thus, in curves of more than about 60°, the Milwaukee brace is not an effective treatment. The components of forces and the moment arms acting correctively in such a curve are small. The total force applied to create any sort of meaningful bending moment would have to be large. An attempt to use pads to exert force of this magnitude usually causes skin breakdown.

In curves of 30° to 60°, the success of the Milwaukee brace may depend on the degree of spinal rotation. Milwaukee braces tend to have maximal therapeutic effect in idiopathic curves of 30° to 60° that lack a significant rotational component. Rotation in a scoliotic spine is not necessarily proportional to the severity of the lateral bend. When rotation is great, pushing on the ribs probably only increases the rotational deformity.

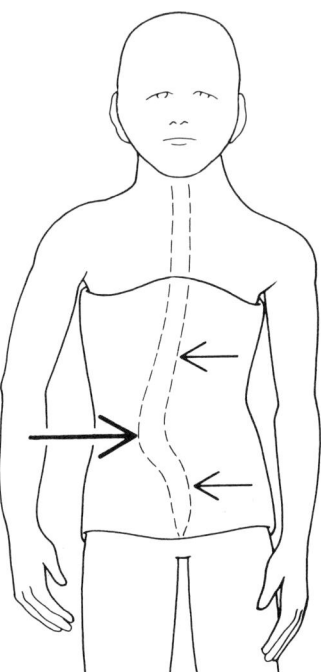

Fig. 1-53. A body jacket, which entirely contains an area of the body, prevents bending forces created by the patient's weight from further deforming the spine.

When treating curves whose apex is relatively low, three-point fixation can be achieved by a molded plastic corset (Fig. 1-55). This thoracal lumbosacral orthosis (TLSO) or Boston brace incorporates all corrective components into a single molded plastic module, which includes a pelvic girdle, an apical convex pad, a left proximal extension, and a void on the concave side to allow for correction.

Just as the Milwaukee brace can be effective in scoliotic curves below 60°, its use can be extended to correct kyphosis. Here, to unbend the curve, forces must be applied anteriorly and posteriorly, ideally directly over the ends and apex of the curves. To unbend the spine, the brace's pelvic girdle can apply a force from its anterior surface. A pad in the thoracic region over the sternum can also be used. To ensure correction and to avoid moving the

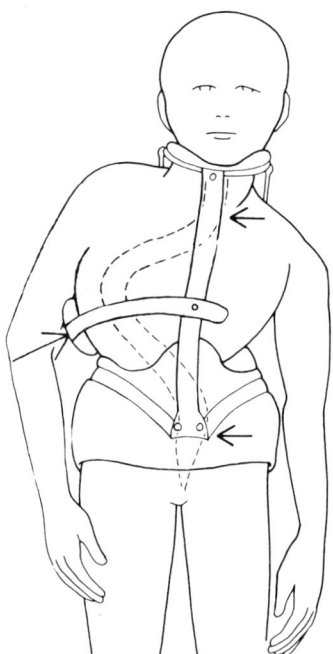

Fig. 1-54. The greater the degree of scoliotic curvature, the more parallel the ribs are aligned with a vertical axis. If a pressure pad is to transmit force to the spine through these ribs, the force applied through the pad must become more vertical. If it does not, it would only squeeze the ribs in toward the center of the body. With greater degrees of scoliosis, a limit to the verticality of the applied force is reached, since in the ultimate position (vertical) no force could be transmitted to the ribs; such a force applied externally would merely push the skin upward.

spine posteriorly, a force acting from the posterior surface directed anteriorly is needed. Such a force can be provided directly over the spinous processes with a pad at the apex of the curve. Such three-point fixation applied with sufficient force can achieve the desired result. It appears that the pads act much as in scoliosis; the discomfort of the brace causes the patient to pull away from the pads, correcting the curve (Fig. 1-56).

Fixation of the pelvis anteriorly, the apex of the curve posteriorly, and the upper sternum anteriorly holds the trunk at three points. Subsequent motion, initiated by muscle activity and trunk movement, produces a relative unbending of the spine. Such activity and motion can be assisted if the weight of the trunk above the apex of the curve can produce a bending force by the shoulder girdle in a direction to correct the kyphosis.

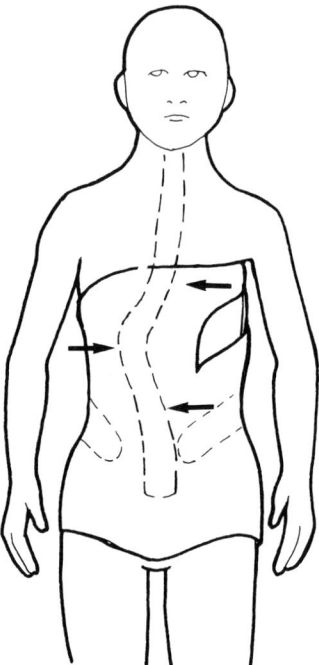

Fig. 1-55. Three-point fixation can be achieved by a molded plastic corset (thoracal lumbosacral orthosis [TLSO]).

16 INTERNAL FIXATION DEVICES

In some cases of scoliosis, external casts and braces do not satisfactorily correct the curve. Surgical correction and internal fixation then become indicated. Surgical correction is achieved by the same principles as with braces and casts by applying proper bending moments. Harrington and Harrington-like instrumentation are methods of straightening the scoliotic spine by implanting parallel compression and distraction rods, which are hooked onto the posterior elements of the spine (Fig. 1-57).

The distraction rods do most of the correction in such systems, much as in halo-pelvic traction or turnbuckle casts. The curve straightens out relatively easily as tension is applied with a distraction rod. As distraction continues, the more the curve is straightened out the more difficult further correction becomes. At first the distraction force has a large component perpendicular to the bent spinal axis and acts through a relatively long lever arm in relation to the apex of the curve (Fig. 1-58). Such a force therefore creates a favorable counter-bending moment. As the spine straightens, the distraction force tends to become parallel to the spine and its lever arm is progressively reduced. Greater force is then necessary to achieve the same

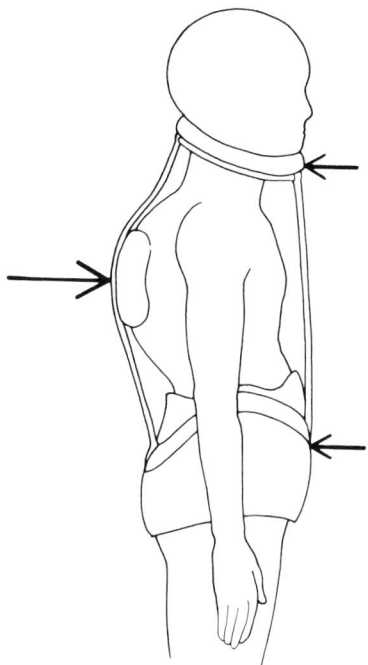

Fig. 1-56. A Milwaukee brace can apply forces in the sagittal rather than frontal plane and thus assists in correcting a kyphotic deformity.

bending moment. Correction ceases when the corrective force equals that of the resistance of the ligaments and discs.

In both theory and practice, a compression rod is less effective than a distraction rod. Compression rods are effective only in curves less than about 50°. In curves greater than that, most of the force derived from the compression hook is parallel to the axis of the spine and acts only through a short lever arm; hence, not much corrective bending moment is created (Fig. 1-59). Furthermore, in curves of more than about 50°, the transverse processes of vertebrae are usually so rotated that the compression rod hooked onto these lies not on the convex side but actually on the concave side (Fig. 1-60). In such a configuration, attempts at compression bend the curve even more. In addition, these transverse processes are pointed more posteriorly than laterally and compression tends to create (or aggravate pre-existing) lordosis (Fig. 1-61).

The strength of the bony attachments into which the hooks of the rods are inserted limit the force that can be applied in either tension or compression. The distraction rod hooks are preferentially placed under the laminae, which, if left intact and not notched, are relatively strong. The compression rod hooks, on the other hand, are put around the transverse processes, which are relatively weak and break if much load is applied.

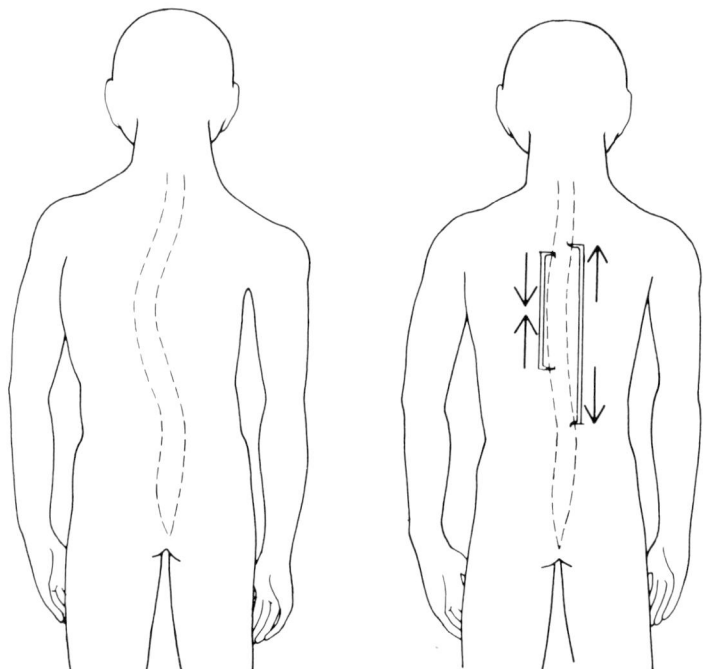

Fig. 1-57. Tensile forces applied in equal and opposite directions to the concave side and compression forces applied to the convex side of the curve create bending moments that can straighten a scoliotic curve.

It is important to ensure that the instrumented spine is balanced over the sacrum. The lowest hook of the distraction rod should be placed vertically over the sacrum to prevent the center of gravity of any overlying residual scoliotic curve from creating bending moments over the sacrum. This would create another curve between the lower end of the rod and the sacrum (Fig. 1-62).

In double curves, it is mechanically advantageous to use one rod in a dollar sign ($) fashion rather than two separate rods. Each curve has its own degree of flexibility, and the chance of straightening the curves equally is unlikely. Therefore, balancing the spine as a whole is more likely with one rod than with two. It is mechanically wiser to accept some limitation in correction and end up with a balanced, stable spine. Furthermore, if two rods that do not overlap are used, a gap between two solid areas acts as an area of stress concentration. The chance of fusion in this inter-rod area is reduced.

Harrington and Harrington-like rods applied posteriorly can correct lordotic deformities associated with scoliosis but are less effective in correcting associated kyphotic deformities. Since rods would have to be bent, the component of force acting to correct the curve is reduced. Also, application of

16. INTERNAL FIXATION DEVICES   **47**

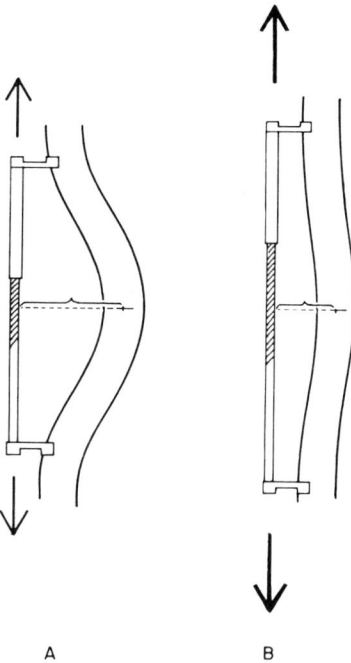

Fig. 1-58. (A) If tensile forces are applied to the concave side of a curve, small forces are needed if the curve is large, since the bending moment arm is also large. **(B)** The straighter the curve becomes, the smaller the bending moment arm. If the same bending moments are to be created, higher tensile forces are needed.

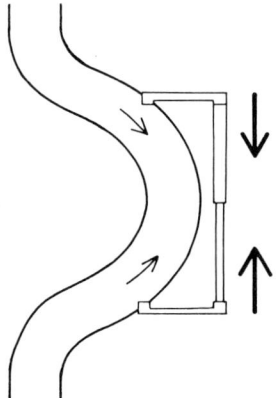

Fig. 1-59. If compression forces are applied to the convex side of the curve, the bending moment created is not very large because the moment arm is small. Large compressive forces would therefore be needed. This is limited by the strength of the bone to which the hooks are applied.

48 PRACTICAL BIOMECHANICS FOR THE ORTHOPEDIC SURGEON

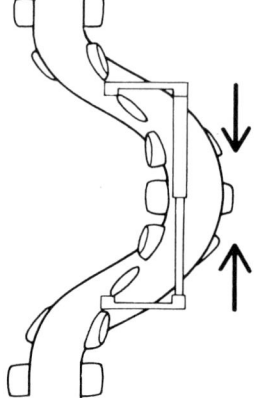

Fig. 1-60. In scoliotic curves greater than 50°, the transverse processes are rotated to such a degree that they are on the concave side of the curve. If compression hooks were applied to these regions, a bending moment would be created that would increase the curvature.

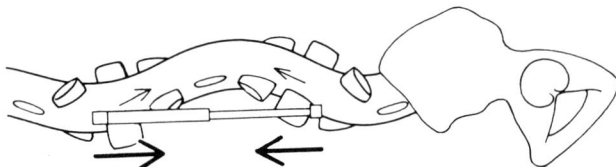

Fig. 1-61. The transverse processes in a severely scoliotic curve on the convex side are rotated posteriorly. If lordosis exists the bending moments created by a compression rod increase the lordosis.

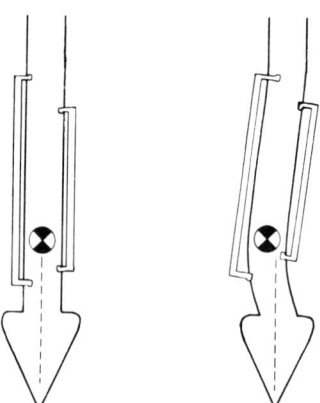

Fig. 1-62. If in correcting a scoliotic curvature the final position achieved is not centered over the pelvis, a bending moment is created proportional to the weight above the pelvis times the amount the center of gravity is off the midline (moment arm).

corrective forces at the ends of the curve, as with Harrington rods, is not mechanically advantageous in short, sharp curves that tend to be rigid in their central portion. It is anatomically impossible to put in Harrington rods if the posterior elements are absent or weak. In such cases another approach is needed.

Dwyer developed an internal fixation device that can be applied to the anterior aspect of the vertebrae. The spine is surgically approached anteriorly and the intervertebral discs are excised. Excising these discs, and bone if necessary, allows considerable correction by removing wedges from the concave side of the curve and thus eliminating part of the spine that contributes to its rigidity. After the discs are removed, the spine is extremely flexible and can be straightened manually. A cable is then applied with staples and screws to the anterior-lateral aspect of the vertebral bodies. The cable holds the correction achieved by the surgical excision of discs (Fig. 1-63) (and bone if necessary); it does not function as a corrective force.

This technique allows the individual correction of one vertebral body on its neighbors as the cable is crimped from interspace to interspace. The Dwyer apparatus, therefore, does not work on the same principles as the methods of treatment outlined above (i.e., by applying a force to unbend a curve). Rather, using the Dwyer method, the surgeon corrects a curve by excising wedges anteriorly, manually straightening out the spine, and then applying a band to resist bending on the tensile side of the corrected curve. This prevents the deformity from recurring. If an attempt is made to use the Dwyer apparatus to correct the deformity (by using it to apply bending moments) rather than just to hold the correction, one risks either pullout

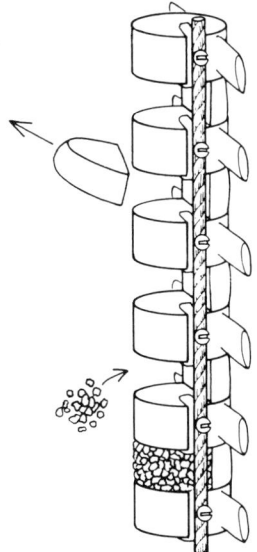

Fig. 1-63. If scoliosis is corrected by the anterior approach, removal of the discs takes away a considerable amount of the internal structure that resists unbending the curve. Little external force is therefore necessary to straighten the spine. The Dwyer apparatus can then be used until fusion occurs to prevent body weight and muscle activity from creating bending stresses that might redeform the curve.

of the fixation of the cable from the bone because of generation of excessive force or tensile fatigue of the cable from excessive repetitive tensile stress.

Newer internal fixation techniques depend less on the distraction/compression approach and more on correcting the associated rotatory deformity inherent in all scoliotic curves. They rely on the fact that when you bend the spinal column, it also rotates. This is because the thoracic spinal segments are attached to the rib cage, which will not allow vertebral bending without rotation. Thus it is possible to unbend spinal curvatures by derotating them. Such fixation systems (eg, Cotrel-Dubousset and Texas Scottish Rite Hospital) (Fig. 1-64) emphasize correcting the scoliotic deformation by derotation to restore normal spinal contours in the sagittal plane. With these segmentally fixed systems, the need for postoperative external support (bracing) is lessened. A further advantage of these derotation-based systems is that they restore the physiologic thoracic kypohsis (Fig. 1-64) that the Harrington instrumentation (distraction/compression) tends to obliterate. Derotation also improves the scoliotic rib deformity, which has both respiratory and cosmetic advantages.

Following the placement of laminar and/or pedicular hooks (for distraction on the concave side of the curves and for compression on the convex side), a prebent rod is placed through the hooks on the concave side. This rod is then rotated 90° so that the spinal contour is converted from a scoliotic curve in the frontal plane to a more physiologic curve in the sagittal plane. A second rod is placed through the compression hooks on the contralateral side, and the two rods are connected together at both ends to form a rigid

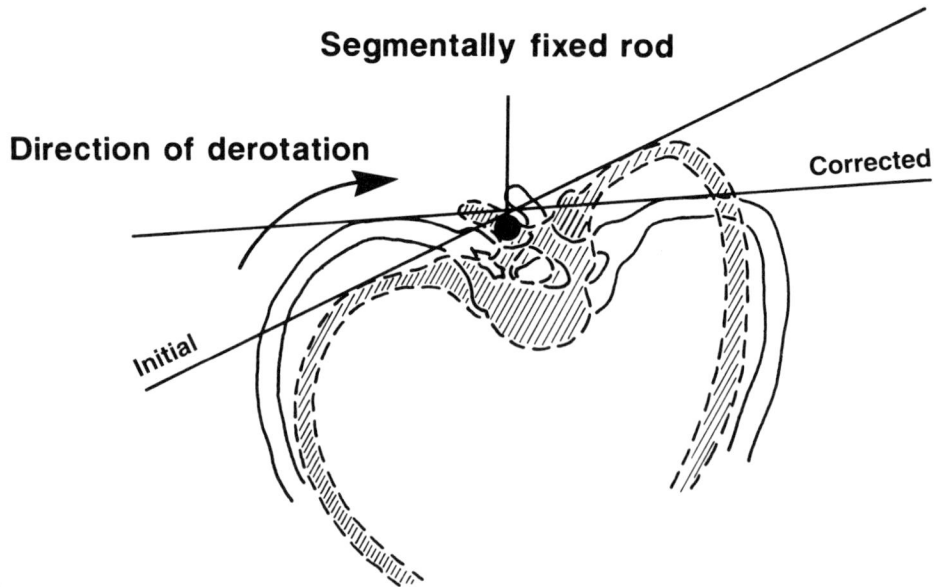

Fig. 1-64. (A) Derotation of the spine is illustrated. The vertebral bodies are rotated in the direction of the arrow, thus lifting the concave depression and depressing the convex rib pump. *(Figure continues.)*

Pre-op   Post-op

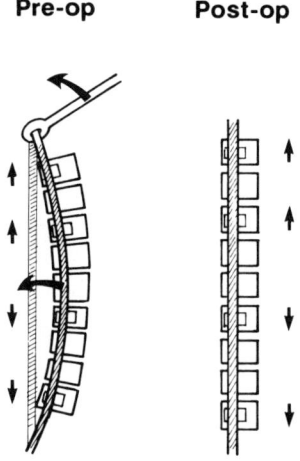

Frontal

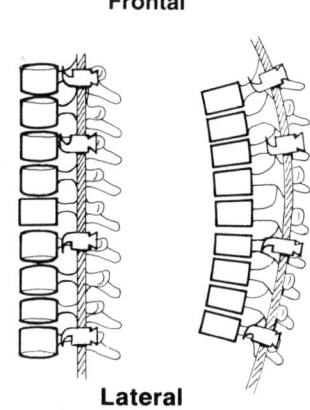

B   Lateral

Fig. 1-64. (*Continued*) **(B)** Schematic right thoracic scoliosis with concave side hook attachments indicated. (Top) Derotation of the concave rod corrects the scoliosis. (Bottom) This process re-creates the physiologic kyphosis. The rod rotates about the fixed end hooks. The apical intermediate hooks pull the concave depression in a posterior direction to derotate the deformity.

quadrilateral. This construct is stable to flexion, bending, and torsion. The stability is sufficient to obviate the need for postoperative casting or bracing. An osseous fusion is used to maintain long-term correction. The Cotrel-Dubosset system employs a combination of forces to unbend and derotate the curve (distraction, compression, apical pressure) through multiple points of fixation between the rod and spinal column.

2

Mechanics of Fracture and Fracture Fixation

1 MECHANICS OF FRACTURE

In general, when a bone is subjected to a steady load, two competing processes occur:

1. *Plastic or viscous flow*, in which planes of atoms or molecules slide over each other like a deck of cards (such deformation is caused exclusively by *shear stress*)
2. *Fracture*, in which a crack grows from microscopic to large size (sometimes rapidly!); for steady loads, fracture in strong, hard materials such as bone is caused by *tensile stress*

Fracture is common in the long bones, but the origins of the necessary tensile stresses are not clear. For instance, to cause tensile stresses by putting the skeleton in traction would generally result in joint dislocation. Also, muscular contractions always resist such tendencies. The tensile stresses that cause fracture generally, therefore, are not caused by tensile loads (or traction) but rather by bending and torsion. These two phenomena are introduced in Chapter 1, Sections 8 and 10, and will be discussed again in this chapter.

Whether significant plastic flow ever occurs in mature bone is controversial. "Greenstick" fracture is sometimes cited as an example of plastic flow in bone. However, greensticking could well represent a combination of small, incomplete cracks or microfracture of one cortex of an immature, poorly calcified bone that has a low modulus of elasticity. Certainly, in the laboratory, conditions can be created that can make bones plastically flow, but whether such conditions occur physiologically with any significant frequency is not known.

2 TENSILE STRESSES IN THE LONG BONES: BENDING AND TORSION

Significant stresses on bones are generated just by the activities of daily living. Consider stair climbing; body weight causes stresses on the bones of the legs as we propel ourselves upward against gravity. The force to move the body weight up the stairs is provided by muscle contraction. The bones are stressed as the muscles contract, bringing their origins and insertions closer together to move the joints. Thus the bony skeleton of the limbs is subjected to asymmetric compression. Bending is the result. Any eccentric or off-center load creates bending, and thus the tibia, femur, and fibula are subjected to bending stress when we go up stairs. The same is true when we walk. Additional asymmetric forces are caused by the relative position of the limbs with regard to the body weight (Fig. 2-1).

Even without motion, double leg stance creates a bending stress in the lower extremity as the body weight is asymmetrically placed relative to the leg. The point at which body weight is centered (at which the body would balance on a pin) is called the *center of gravity*. In the human, the center of gravity is just in front of the second sacral vertebra.

It has been calculated that the stresses during walking generated by the muscles on the hip are sufficient to bend the femoral neck permanently. Similar calculations can be made for the long bones of the upper extremity. Enormous forces can build up with isometric contractions, as in using one

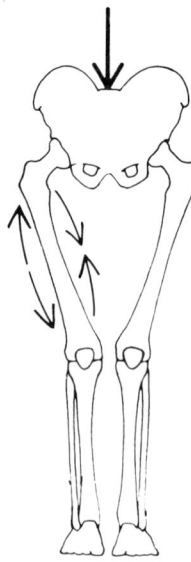

Fig. 2-1. Asymmetric loading on the femur causes bending stresses.

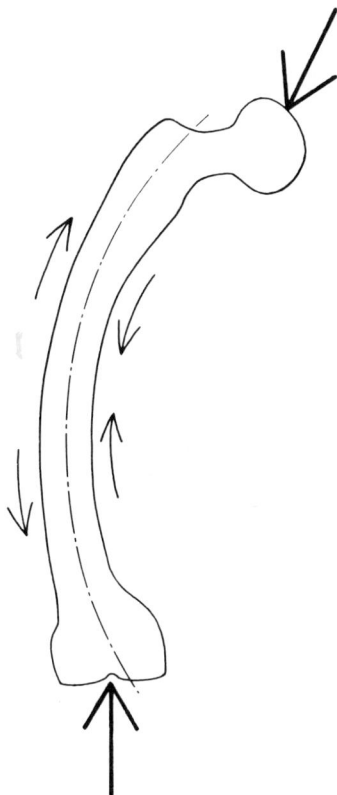

Fig. 2-2. Bending stresses are tensile on the convex side and zero at the center line (neutral axis). (Strictly speaking, the neutral axis is a "center of gravity" and need not be at the center line.)

hand to oppose a forceful push by the other. Diseased bone frequently fractures, and normal bone subjected to impact loads can fracture, but the amazing thing is that normal bone does not continually fracture even though it is subjected to rather substantial bending stresses brought about by the activities of daily living. We now discuss why such disasters are rare.

Figure 2-2 illustrates the tensile and compressive stresses caused by long bone bending. As discussed in Chapter 1, a neutral axis can be defined for any structure subject to bending. The bending causes compression on the "concave" side of the neutral axis and tension on the "convex" side. The neutral axis is where the stress state changes from compressive to tensile and the stress is, at that axis only, zero. Thus there is an important tensile component in bending. In a solid such as bone, tension is a more potentially destructive stress than is compression. It is tensile stress that initiates fracture in bending (Fig. 2-3) and in most other common situations as well.

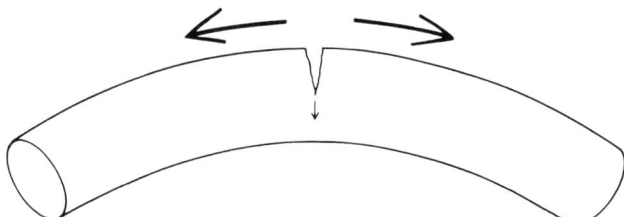

Fig. 2-3. Fracture in bending always begins on the tensile side.

Many factors contribute to the resistance of bone to bending and keep the skeleton intact during normal physical activity. An uncontrolled or substantial fall or high velocity impact is generally necessary to break a normal bone.

One of the primary mechanisms for stress relief operating in the appendicular skeleton is composed of the articulations. Instead of our limbs bending in midshaft, most of the bending takes place at the joints, like articulations in the roadway of a bridge. Rather than the relatively rigid roadway bending, the articulations rotate (Fig. 2-4).

Muscles also function to reduce bending stresses in bone. They act as "guy wires" to reduce bending. (Guy wires are familiar devices used to hold up high antennas and telephone poles.) In performing this function, the muscles increase the compressive stress in the bone. This is not a disadvantage, since bone, like most hard materials, has a greater resistance to fracture in compression than in tension. Muscles that straddle joints reduce bending stress by acting as guy wires and supporting part of the bones so that the entire bone is not subjected to bending (Fig. 2-5).

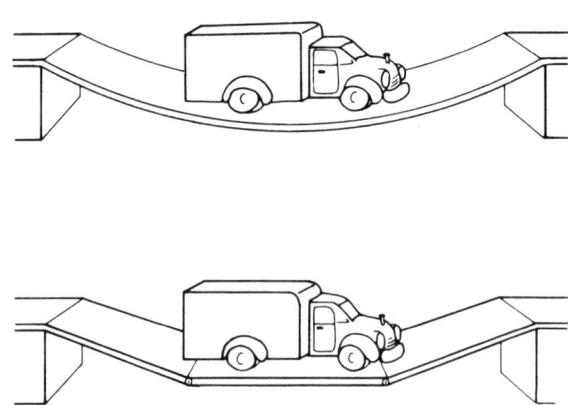

Fig. 2-4. Articulations reduce bending in each section of the road.

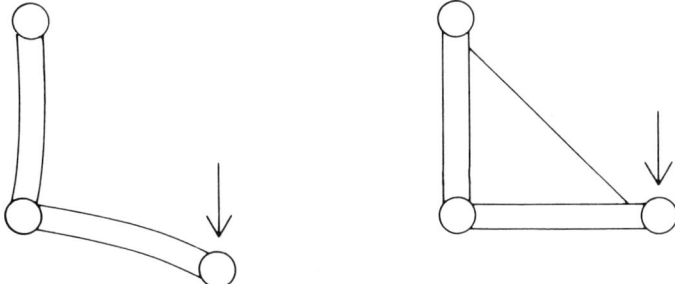

Fig. 2-5. Muscle force introduces stresses opposite to the load, canceling much of the bending stresses and reducing the overall stress.

Biaxial muscles, which cross two joints, are even more effective in reducing the bending stress on bone. An added advantage of biaxial muscles is that most activities of daily living require combinations of joint actions, such as dorsiflexion of the wrist with flexion of the fingers or plantar flexion of the ankle with flexion of the knee. Such arrangements of biaxial muscles make maximal use of them and help, in part, to explain the functional advantages of the rather complicated arrangement of so many muscles on the appendicular skeleton.

The structure of bone itself is designed to minimize bending stress. Bones are curved to be in line with the predominating resultant force that acts on them, increasing their compressive stress but decreasing their tendency to be bent (with its concomitant tension) (Fig. 2-6)—again, a trade-off of tension for compression.

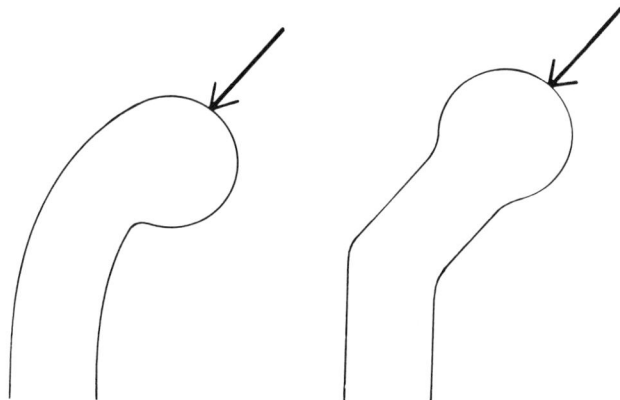

Fig. 2-6. Each shape reduces bending at the end of the bone. The force is now lined up with the neutral axis and has no bending moment in the epiphyseal region.

As Figure 2-2 indicates, the stresses are lowest near and at the neutral axis and increase with distance from it. The maximum values for both tension and compression occur on the outside of the bone, at the *extreme fibers*. The material at the extreme fibers also has the longest lever arm with regard to the bending moment and therefore is most important in resisting bending. The hollow tubular structure of bone is extremely effective in maximizing flexural strength and rigidity; the material that contributes least to these properties would be near the neutral axis and is eliminated (see Chapter 1, Section 8). In general, hollow, tubular shapes provide maximum resistance to bending and twisting (flexural and torsional rigidity or section modulus), with minimal weight. Although bones are irregular in shape, the neutral axis can be located with some effort. Once the neutral axis has been located the relative resistance to bending in a particular plane can be calculated. Thus each bone has a resistance to bending that increases with its mass out at the extreme fiber or, in other words, the mass multiplied by its lever arm to the neutral axis. This can be quantitated and is referred to as the *area moment of inertia*. Each bone, along its predominant bending axis, has an area moment of inertia that depends on its geometry and distribution of mass relative to that geometry.*

The tubular cross section of long bones facilitates circulation, repair, and nutrition, because two surfaces with osteogenic potency are available rather than one and the hard tissue thickness that must be penetrated by osteoclasts, Haversian canals, and associated blood vessels is much reduced by the tubular shape. All activities such as standing, walking, carrying, throwing, and pounding produce a predominant tensile stress on the convex side of the long bones as bending stresses are created within the bone. In gait, at foot flat, the maximum tensile stress on the tibia is posterior. The predominant tensile side of the femur is lateral, because it deviates back into the body from the intertrochanteric area, so that the knees almost touch (Fig. 2-1). All the large bones are acted on as levers by many muscle groups, and levers are always subjected to bending. Bending is the predominant stress in the bones of the upper extremity as well. In the act of throwing, pounding, lifting, or carrying the predominant tensile side of the humerus and forearm bones is posterior (Fig. 2-7). As shown later in this chapter, knowledge of the predominant tensile and compressive side of long bone is important in considering optimal placement of internal fixation devices and bone grafts.

It has been shown by placing strain gauges on human and animal bones during gait and stance that the bones do bend and that the bending strains

---

*In gross approximation, the material in a bone resists bending as the cube of its distance from the neutral axis. This value gives some idea of the tremendous difference in strength achieved by even a small increment of bony material further out from the neutral axis.

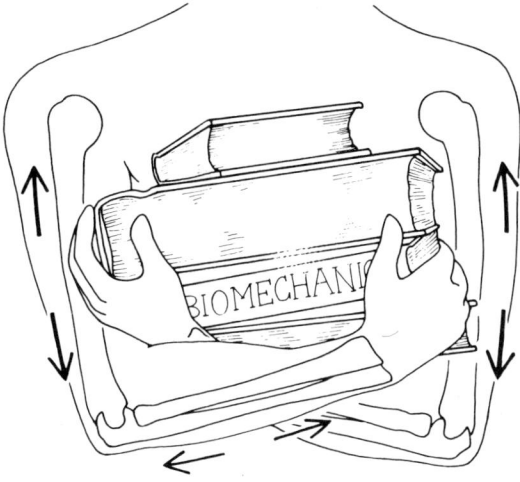

Fig. 2-7. The posterior humeral, ulnar, and radial surfaces will tend to have the highest tensile stresses in most situations.

created are physiologic. Long bones are predominantly subjected to bending stress, but are also subjected to compression and torsion. Consider the rotation of the tibia during normal gait. As in bending, there is an important tensile component in torsion, as the next section shows. As in bending, there is a neutral axis in torsion and the resistance to torsion depends on the distance of the extreme fiber from its neutral axis. Again, this peripheral material has the greatest lever arm, and its distribution can be quantitated. This value for torsional resistance, known as the *polar moment of inertia*, for long bones approximates the fourth power of the distance of the material from the neutral axis. Thus the hollow structure of long bones is maximally effective in reducing torsion as well as bending with a minimum investment of bony structure.

3 MECHANICS OF FRACTURE: TENSILE STRESS AND STRESS CONCENTRATIONS

Fracture is very much a matter of the distribution of stress and *mechanical energy*. For instance, the work necessary to fracture the average human tibia is only about 1/10,000 of the kinetic energy of an 80 kg skier at 10 m/sec (24 mph). Disaster occurs only when the kinetic energy is abruptly (and painfully) concentrated and converted to the work necessary to strain the tibia; even then, as we shall see, certain types of strain are far more harmful than others.

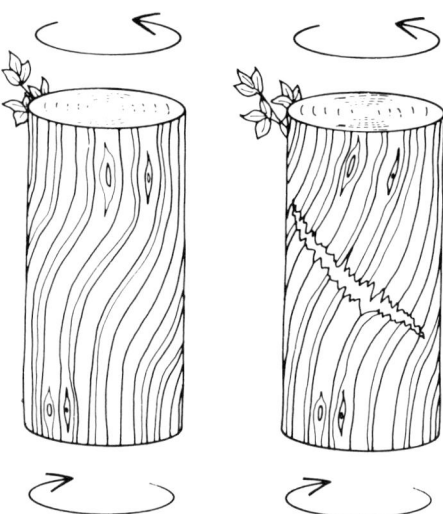

Fig. 2-8. The highest tensile stress is on a plane 45° from the highest shear stress. In torsion the fracture crack tends to follow a spiral plane to maintain this angle.

The stresses that occur in bending are discussed in Chapter 1. In bending, as Figure 2-3 shows, the fracture begins at the convex surface, at the extreme fiber with the highest tensile stress. If any grooves, scratches, or other such features are on the extreme fiber, the crack begins at that point. In any case, the crack proceeds across the structure, perpendicular to the tensile stresses caused by the bending (Fig. 2-3). Transverse fractures of long bones are obviously due to bending. However, spiral fractures are due to torsion or twisting forces.

As shown in Figure 2-8, twisting or applying torsion to a log of wood subjects it to *shear strains* (and stresses) in the horizontal plane. As twisting proceeds, the longitudinal fibers are stretched by the distortions caused by the shearing. The fibers also shear in a nearly vertical direction, but neither horizontal nor vertical shearing is as critical as the stretching or tensile deformation. It is possible to show that the tensile stress is maximal at a 45° angle to the axis of torsion, and, as shown in Figure 2-8, fracture will occur at this angle. The fracture crack in many cases continues at this angle and thus describes a 45° helix. The result is a "spiral" fracture.

In most rigid materials, then, tensile stresses cause transverse fractures when bending forces are applied and spiral fractures when torsional forces are applied. The exceptions to this rule are for very *anisotropic* materials, such as wood, in which the structure and the properties are highly directional. Bone, similar to wood in this respect, has planes of strength and planes of weakness; any long bone is much stronger in tension along its shaft than in a transverse or tangential direction.

3. MECHANICS OF FRACTURE: TENSILE STRESS AND STRESS CONCENTRATIONS

Cases of true bending or torsion are relatively simple. There is always, at any point, a maximum shear stress on one plane and a maximum tensile stress on some other plane. In some cases where there is little tensile stress but considerable shear stress, fracture is inhibited and plastic flow first occurs. For instance, bending or torsion may be carried out in a fluid under large pressures; net tensile stresses may not even exist in such cases. Even solids with a high resistance to plastic flow (such as extremely *brittle* materials), which ordinarily shatter without deforming (such as glass), can deform under such circumstances.

On the other hand, in some circumstances tensile stresses are high and shear stresses low. Then, even materials that normally deform easily without breaking—*ductile* materials—can fracture catastrophically. Thus it is always important, in any mechanical situation involving failure, to assess the relative magnitudes of shear and tensile stresses.

Tensile stresses can be quite high at a *stress concentration*. Consider Figure 2-9, in which the cross section of a bar changes suddenly; the bar is loaded so as to be in tension. In mechanical equilibrium the total load is

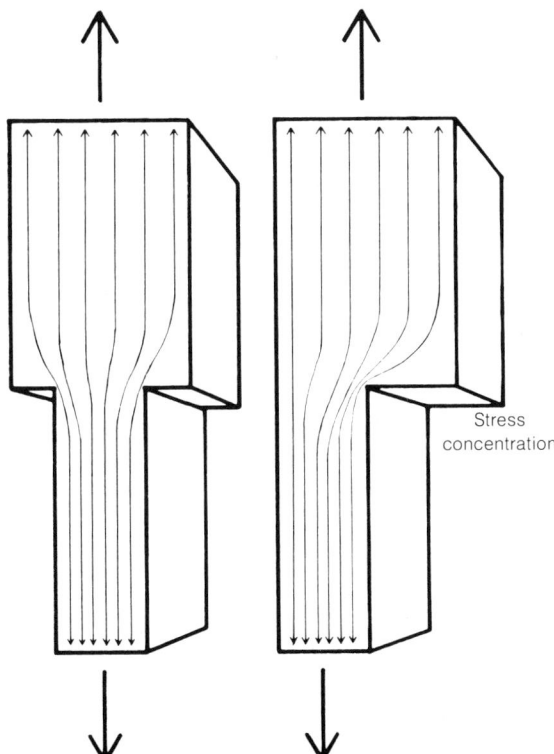

Fig. 2-9. Stress is concentrated by sudden change in cross section, at sharp interior corners.

the same for any cross section, so the stress must be higher in the narrow end of the bar, in inverse proportion to the ratio of the cross-sectional areas. To represent the stress, we use lines of force in Figure 2-9; so many lines per cross-sectional area correspond to so many newtons per square meter, and the total number of lines is proportional to the total load on the bar. Thus, for mechanical equilibrium, each cross section must contain the same number of lines. A remarkable situation in the immediate vicinity of the change in section is a great concentration of stress, as shown by the concentration of the density of lines. Although the exact computation of the stress concentration requires use of the mathematical theory describing elasticity, stress concentrations can be measured by a variety of techniques and have been calculated for a number of common cases such as holes or notches in plates.

In Figure 2-10 we show two ubiquitous geometric figures in which large stress concentrations occur. Holes, notches, grooves, threads, keyways, any change in section—all these (and many more!) serve to concentrate the

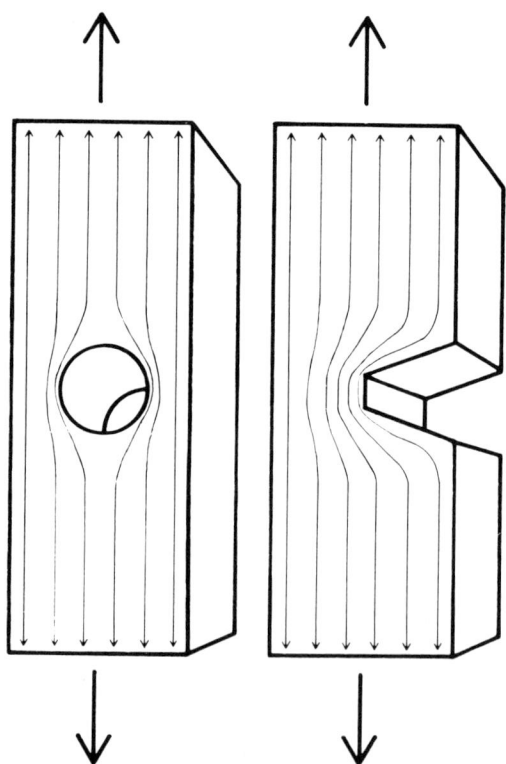

Fig. 2-10. Stress is concentrated at the equator of the hole and at the bottom of the notch. A sharp notch would concentrate the stress further.

3. MECHANICS OF FRACTURE: TENSILE STRESS AND STRESS CONCENTRATIONS

stress. In a sharp, deep crack, tensile stresses may be concentrated by a factor of 10,000! As it turns out, solid materials of any kind contain a huge variety of microscopic defects, scratches, cracks, pores, and so forth, that can have sharp tips. Brittle materials (eg, glass) are readily cut by scribing a sharp scratch on the surface and then loading so that there is a tensile stress across the scratch. If such a scratch exists at the bottom of a thread or at a hole or cross-sectional change where the stress is concentrated, fracture may occur even at low loads.

Because stress concentrations and some kinds of defects are unavoidable, in most cases solid structures would be much weaker than they are if the stress at the tip of a crack or scratch were the only factor that determined the likelihood of failure. Even if the local stress concentration were millions of newtons per square inch, enough to rend apart the chemical bonds at the crack tip, the crack would not progress unless there was enough energy to push it along. This energy is necessary because fracturing of material creates new surfaces, and creation of new surface means the rupturing of chemical bonds, which requires energy; per unit area, so many ergs (or calories or BTUs) of energy are needed. This quantity is called the *surface energy* or surface tension.

Applied forces supply necessary energy for the creation of new surfaces. New surface is created by growth of the crack. The occurrence of fracture then depends on a balance between available mechanical energy and energy needed to make the crack grow. The fracture stress for crack growth is also related to initial depth of surface crack or scratch. Deeper cracks weaken the material more drastically; the square of the stress required for fracture is inversely proportional to the crack depth. Thus doubling the crack depth decreases fracture strength by nearly 30%. On the other hand, increased stiffness (modulus of elasticity)* or surface energy increases the resistance to fracture.

In these discussions we have assumed that the applied tensile stress is perpendicular to the crack. As Figure 2-11 shows, if the tensile stress is parallel to the crack, the crack does not spread. (The spreading of the crack has no effect on the parallel case; we should end up with two fragments with no change in stress.) If the crack is at some angle to the direction of the tensile stress, then we should consider the component of the stress that is perpendicular to the crack.

As expected, larger cracks are "weaker." As any object contains an assortment of cracks or potential cracks (eg, lacunae, canaliculae, Haversian canals, and cement lines in bone), fracture is a matter of the "weakest link"

---

* The general relation for fracture of a brittle material is

$$(\text{fracture stress})^2 = \frac{(4 \times \text{modulus of elasticity} \times \text{surface energy})}{(\text{depth of crack})}$$

64 PRACTICAL BIOMECHANICS FOR THE ORTHOPEDIC SURGEON

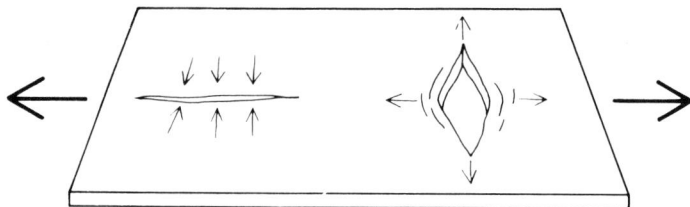

Fig. 2-11. Crack perpendicular to the tensile stress opens and spreads, leading to fracture if the stress is sufficient.

(ie, the largest crack giving way). Compressive stresses should not encourage the growth of cracks, but rather retard such processes.

4 ENERGETICS OF FRACTURE, FRACTURE TOUGHNESS, AND IMPACT

The preceding discussion really describes the behavior of brittle materials. By "brittle" we mean a material that does not deform plastically or by viscous flow before fracture. Typically, very hard materials are intrinsically brittle; they do not deform until high stresses (well above the fracture stress) are reached.

Suppose the stress versus strain behavior of two materials is plotted, as in Figure 2-12. The work (per unit volume) to fracture each material is the total area under its tensile curve. Thus the soft, annealed copper requires

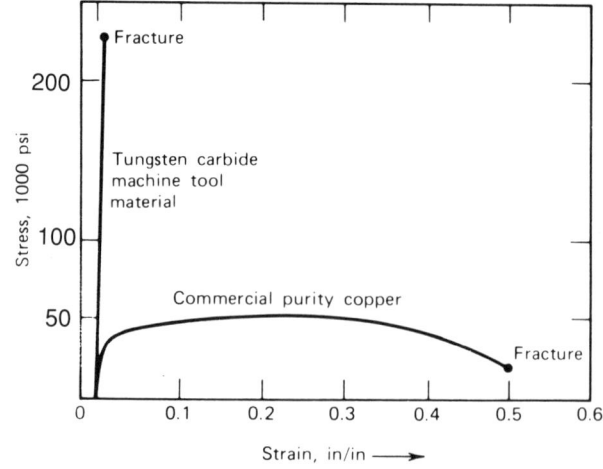

Fig. 2-12. Stress versus strain for a soft, ductile material (copper) and a hard, brittle material (tungsten carbide).

much more work to fracture it than the extremely hard carbide tool material. The work required to fracture a material is referred to as fracture *toughness*.

One reason why softer materials may be tougher is demonstrated by what happens at the tip of a crack under stress in such a material (eg, stainless steel). The large stresses at the crack tip cause local viscous or plastic flow. As a result, more work must be absorbed by the crack; the material near the crack is severely deformed. This work must be added to the surface energy. Thus ductility, however small, increases the fracture stress. A sudden impact leaves little or no time for plastic or viscous flow and thereby causes failure much more readily. The effect is further amplified because a fracture crack accelerates if the energy to do so is available; however, once acceleration occurs, less plastic flow occurs and therefore less energy is needed to propagate the crack and further acceleration is inevitable. On the other hand, if for some reason the crack slows down, more energy is needed to continue, and the crack decelerates until it stops.

5 FATIGUE FRACTURE; "MARCH" FRACTURES; RESISTANCE OF CORTICAL BONE TO FRACTURES

Even at low stress levels, far below the stresses necessary to cause catastrophic failure or observable viscous or plastic flow, the seeds of failure may be sown. The most common type of physiologic loading is cyclic or

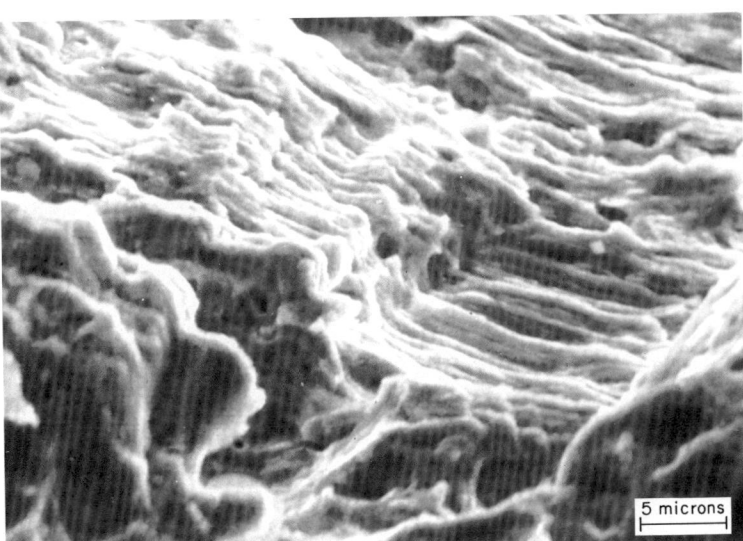

Fig. 2-13. Fatigue striations on the fracture surface of a fractured Schneider pin. The striations are 0.25 to 0.50 μm wide (1 μm = 10^4 cm).

Fig. 2-14. Striations (less than 1 μm wide) produced in a fatigue testing machine programmed to produce seven cycles at high stress and then seven cycles at low stress, repeatedly. The striations are therefore in groups of seven, wide and narrow, since higher stress produces larger striations. (Courtesy R.M.N. Pelloux.)

intermittent; locomotion puts repetitive or cyclic stresses on the lower extremities. Even if the stress is well below the fracture stress so that a preexisting crack does not immediately grow, there may be enough to advance it a few microns. Each time the stress is reapplied, the crack advances another few microns. Eventually the crack is large enough to increase catastrophically. Such a phenomenon is called "fatigue failure." We have illustrated this process in Figure 2-13. A scanning electron micrograph shows the markings left by the crack; each "fatigue striation" in the micrograph marks the place where the crack stopped and then resumed its growth.

The larger the intermittent or cyclic stress, the farther the crack advances with each application, as shown in Figure 2-14. Thus the larger the cyclic stresses, the faster the rate of growth, and the final failure occurs in a shorter time (ie, the "fatigue life" is shorter).

In Figure 2-15 we present typical fatigue data. The number of cycles to cause failure is plotted as a function of the maximum stress applied during each cycle. Variation of stress with time is also illustrated.

For many materials there is a stress level below which the fatigue life is practically infinite. This stress level is called the *fatigue limit* and is indi-

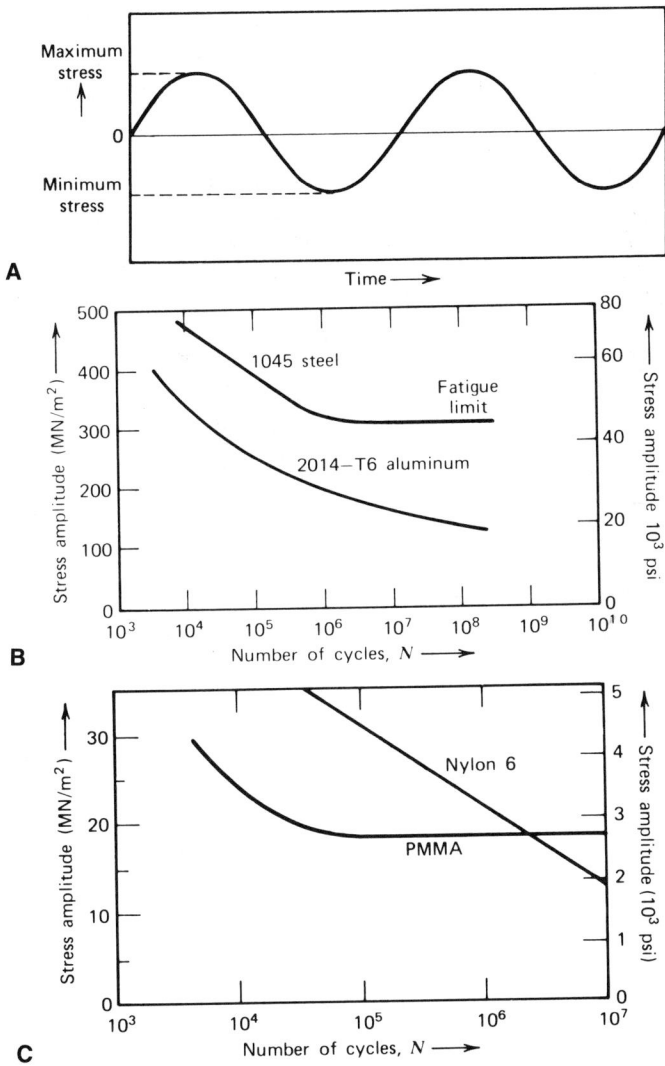

Fig. 2-15. (A) Time-varying loading cycle that is often used to determine the fatigue test. Maximum tensile stress reached during one portion of the cycle is equal to the maximum compressive stress reached during the other portion of the cycle when the load is reversed. **(B)** Experimentally determined stress versus number of cycle curves for a plain-carbon steel (0.47% carbon) and an age-hardened aluminum alloy. These curves are for fully reversed stresses (see Fig. 2-15A). The steel shows a fatigue limit (ie, a stress below which it will not fail regardless of the number of cycles). The aluminum alloy shows no fatigue limit, but the slope of stress versus log N becomes less negative as the stress decreases. (Adapted from Hayden HW, Moffatt WG, Wulff J: The Structure and Properties of Materials. Vol 3: Mechanical Behavior. John Wiley & Sons, New York, 1965, with permission.) **(C)** Stress versus number of cycle curves for the polymers nylon and PMMA (Lucite). Lucite shows a fatigue limit, but nylon does not. (From Riddell MN, Koo GP, O'Toole JL: Polymer Eng Sci 6:363, 1966, with permission.)

68 PRACTICAL BIOMECHANICS FOR THE ORTHOPEDIC SURGEON

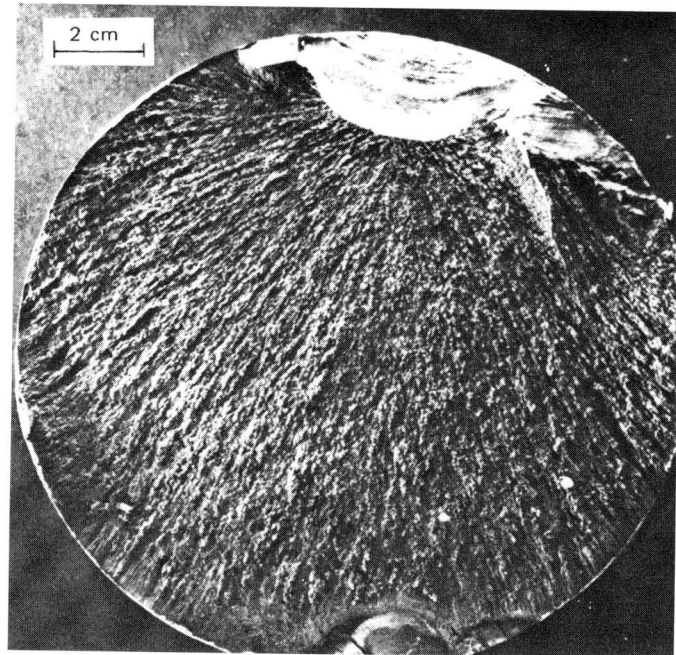

Fig. 2-16. Macrostructure of a fatigue failure surface of a steel piston rod. Fracture originated at the top edge. The smooth area, with the "clamshell" markings, corresponds to slow fatigue crack growth, and the dull fibrous section is the region of fast fracture. (From American Society for Metals: Metals Handbook. Vol 9. Ed 8, 1974, with permission.)

cated in Figure 2-15B for steel. Obviously the fatigue limit is a good level to stay below in any mechanical design. However, if stress concentrations are present, this may be impossible. This, then, is by far the most common history of mechanical failure: first, a stress concentration occurs, through defective design, defective materials, accidents, or careless handling; from

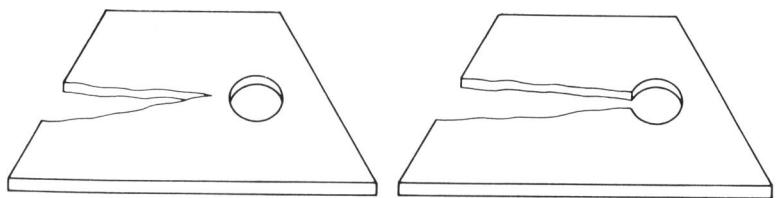

Fig. 2-17. Advancing crack blunted and stopped by a hole.

Fig. 2-18. Haversian systems may resist fracture in the same way as in Figure 2-17.

the stress concentration, a fatigue crack slowly grows; when the fatigue crack has grown large enough, the final catastrophic failure occurs. It is often possible to determine this course of events with the naked eye by examining the fracture surfaces (Fig. 2-16), because the textures left by fatigue and catastrophic propagation are different.

Fatigue failure of bone occurs; "march" fractures are a good example. Bone with a higher density of Haversian systems has a greater fatigue resistance because the cement lines and Haversian canals serve to divert and arrest crack growth. The subdivision of bone by cement lines tends to guide the crack longitudinally; however, if the osteons tend to bend individually, there is less rigidity in bending.

The Haversian canals also help to prevent fracture, as shown in Figures 2-17 and 2-18. The canal has a much larger radius of curvature (ie, it is not as "sharp" as the crack edge). Therefore, much higher stresses are required to resume crack propagation. (In the early days of aviation, it is said that the precaution against fatigue failure of the wing was to drill holes along the probable path of the crack!) For march fractures, another kind of fatigue is likely important (ie, fatigue of muscles that act to reduce tensile stresses caused by bending of the long bones).

The ultrastructure of bone is also remarkable in its effect on mechanical properties. On the finest scale of observation (electron microscope), bone consists of a highly ordered protein (collagen) matrix reinforced with mineral (hydroxyapatite, $Ca_{10}[PO_4]_6[OH]_2$) crystals. Collagen is known to be a relatively soft, pliable material; the mechanical properties of soft connective tissue typify collagen. Bone mineral, as observed in vitro, is brittle and friable, resembling chalk in mechanical properties. Neither the collagen nor the mineral would serve satisfactorily as a skeletal structural material by itself: collagen is not stiff enough, and the mineral is too brittle. However, the composite material, with fine crystals of mineral (about 0.05 μm long and 0.005 μm wide) embedded in the protein matrix has excellent mechanical properties, similar to those of teakwood. (In fact, the properties of bone are superior in that there is less directionality or "grain.")

6 CORROSION OF METALLIC IMPLANTS

Many commercial alloys with extremely high strength would be useful in fixation devices and prostheses, but are unacceptable because of inadequate *corrosion* resistance. Corrosion products can, in the extreme, cause bone necrosis and "rust granulomas" in the adjacent soft tissues; in less severe cases, pain and inflammation may occur under aseptic conditions. Many designs and materials now used are marginal with regard to corrosion resistance. Care should be exercised in their use, as discussed below.

As Figure 2-19 shows, the driving force for corrosion is also the basis of the electrical storage battery. The energy released by a chemical reaction may be used to drive electrons through a circuit or device and do useful work, but only at the cost of exhausting the battery by partially or completely consuming one of the reactants. The consumption, when it occurs unintentionally, is commonly referred to as corrosion. If an iron nail is

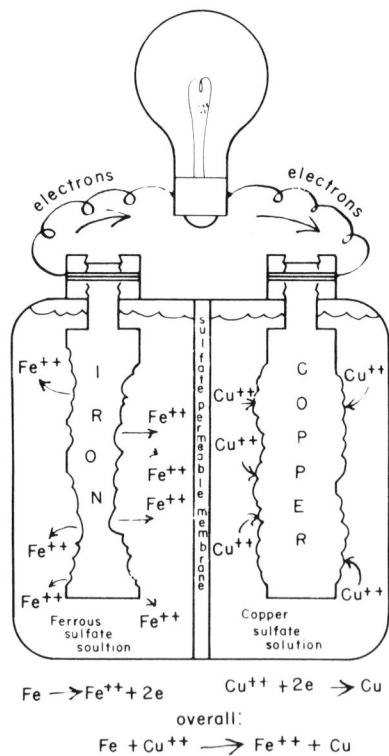

Fig. 2-19. Battery based on the reaction $Fe + Cu^{++} \rightarrow Fe^{++} + Cu$. The iron corrodes and is ultimately dissolved.

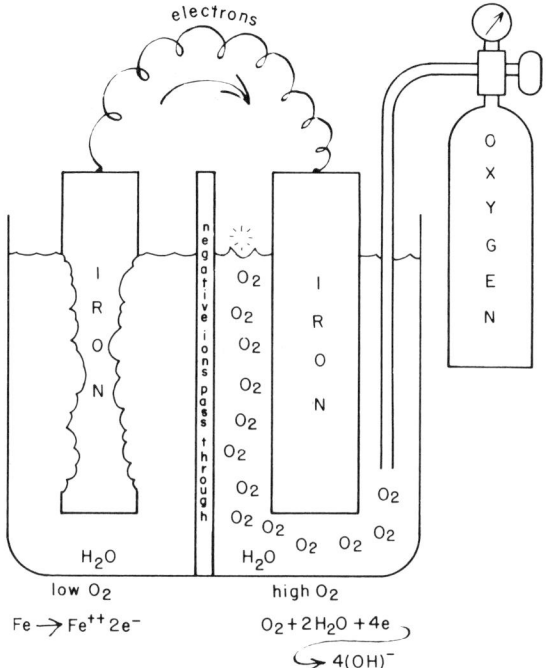

Fig. 2-20. Battery based on a difference in oxygen concentration. The iron electrode in the low oxygen side corrodes.

dipped into copper sulfate solution a reddish deposit appears on the nail. The deposit is copper, which is displaced from the solution by the reaction

$$Fe + Cu^{++} \rightarrow Fe^{++} + Cu.$$

In general, a more active metal always displaces a less active (or more "noble") metal from solution. We could, if we wanted, arrange to have the electrons transferred from the iron to the cupric ion through a wire rather than directly. The resulting device would be a battery (with a maximum voltage of about 0.75 V).

Thus corrosion occurs because of differences in chemical reactivity, which gives rise to electrical currents generated by destruction of the more reactive material. The example used in Figure 2-19 is based on reactivity differences caused by dissimilar metals. However, even the same metal has different reactivity in different environments. A common environmental difference that can cause corrosion is a different oxygen concentration. This can occur in a body fluid or in water. In a situation such as that shown in Figure 2-20, with identical iron electrodes but higher concentration of dissolved oxygen on one side, the reduction reaction has greater "reactivity" on the

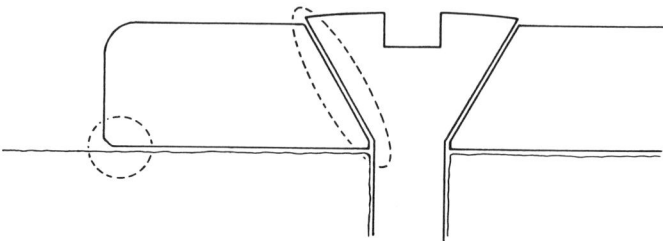

Fig. 2-21. Probable sites for "crevice" corrosion based on the oxygen concentration cell shown in Figure 2-20.

oxygen-rich side. Therefore, the following reaction occurs on the side with higher oxygen:

$$O_2 + 2H_2O + 4e^- \rightarrow 4(OH)^-.$$

The reaction on the "oxygen-poor" side is

$$Fe \rightarrow Fe^{++} + 2e^-.$$

As before, the electrons generated by the oxidation reaction (on the left side in Fig. 2-20) are fed through the electrical connection to the reduction reaction on the right side of the cell in Figure 2-20.

Thus differences in oxygen concentration can lead to corrosion, and the damage occurs where there is least oxygen. The most extensive corrosion occurs underneath the heads of screws and other fasteners. The screw head has free access to the oxygen around it, but just beneath the screw head it is more difficult for oxygen to penetrate. Consider Figure 2-21, which shows the most likely areas of corrosive attack on a bone plate. All areas are exposed to seepage of body fluids, but access to dissolved oxygen is slight. This is crevice corrosion, the most common in surgical implants.

Figure 2-22 is a clinical example. The plate was removed because corrosion under the screw heads and the edges of the plate were causing considerable pain.

There is another dimension to this problem. Stainless steels and other alloys used in orthopedic implants are effective against corrosion because continuous, tightly bound oxide films are on their surfaces. The alloys themselves, if perfectly clean, are quite reactive. This reactivity binds the protective film tightly. Nitric acid makes the films even better. However, chloride ions puncture the film. In the body, Cl^- is readily available, and even with low H^+ levels present, many alloys that are considered very resistant to corrosion become rapidly and extensively pitted because of failure of the protective films.

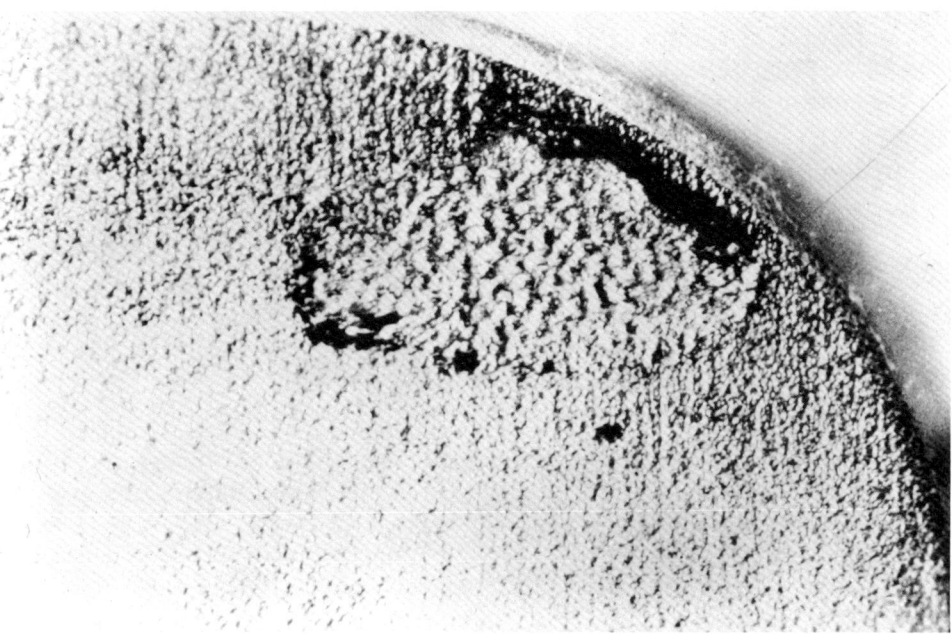

Fig. 2-22. Corrosion at the edge of a bone plate, as shown schematically in Figure 2-21. (Courtesy of J. Cohen.)

Finally, there is some evidence that metal ions liberated at the *anode* portion of the crevice hydrolyze the water in the vicinity. In physiologic media, then, each corroding crevice is a veritable HCl generator!

To make it worse, stress concentrations can occur at crevices, and stress always makes metals more reactive. A stress concentration alone can create an anode and cause corrosion; together with the oxygen concentration difference between the inside and outside of the crevice, the effect is reinforced.

Finally, consider corrosion pits on a previously smooth surface as stress concentrators that can start fatigue cracks. Also, the stress concentration at the crack tip accelerates corrosion, as does the "crevice" or oxygen concentration effect. Obviously, corrosion can drastically accelerate fatigue failure. (In practice, fatigue life in a corrosive environment may fall to 1/1,000 or less of its former value.) Also, the paramount importance of avoiding stress concentrations and optimizing corrosion resistance is apparent. To achieve the latter

- Use resistant alloys
- Avoid dissimilar metals
- Avoid ion concentration difference
- Avoid oxygen concentration differences, including crevices
- Avoid stress concentrations

7 IMPLANT MATERIALS FOR INTERNAL FIXATION DEVICES

In general, the following are straightforward rules for selecting and making an orthopedic implant.

- The material and design should have adequate mechanical strength and fatigue resistance. Since bone has tensile strength of (roughly) 70 million newtons/m^2 (10,000 psi) and a compressive strength of about twice that number, and the size of the implant is restricted, the material should tolerate stresses of 700 million newtons/m^2 (100,000 psi).
- Corrosion resistance should be extremely good.
- There should be no toxic, carcinogenic, or allergenic reaction.

As shown in the next section, the mechanical functions of most implants involve tensile stresses in particular. Therefore, tensile properties are of greatest interest. A familiar engineering test of tensile properties is the tensile test described in Figure 2-23. A specimen is strained at a constant rate and the stress recorded as a function of strain. (Alternatively, the load could be slowly increased and the strain measured as a function of stress.)

Several important features in the tensile test data are shown in Figure 2-24. First, the stress at which permanent plastic deformation occurs is the *yield strength* (or yield stress)—generally taken as the stress at which the specimen is permanently stretched 0.2%. Since the specimen is also stretching elastically, it is necessary to use the elastic (Young's) modulus to subtract the elastic strain from the total strain as the test continues, to identify the yield point. After yielding, the material may, if it is ductile, continue to deform, usually with increasing stress, until a maximum stress, the *ultimate tensile stress*, is reached. This is the second feature of general interest. Finally, when strained sufficiently, the material fractures. The strain or *elongation at fracture* is the third feature of general interest, as it indicates the ductility of the material.

As the tensile curves in Figures 2-12 and 2-24 show, some materials have very large elastic strains. Some are so brittle that they fracture before the yield point occurs; others deform so easily that there is no ultimate tensile stress. However, the metallic alloys used for implants generally exhibit all three features, and their relative values are used to compare the materials. Ideally, all three, the yield stress, ultimate tensile stress, and elongation at fracture, should be as large as possible. In practice, some trade-offs between strength and ductility have to be made. The requirement for high ultimate tensile strength (about 700 million newtons/m^2) has in the past restricted the choice of structural materials for fixation and prosthetic devices to metallic alloys. Future possibilities such as strong ceramics and polymeric fiber composites are being investigated. At present five metallic alloys are in general use.

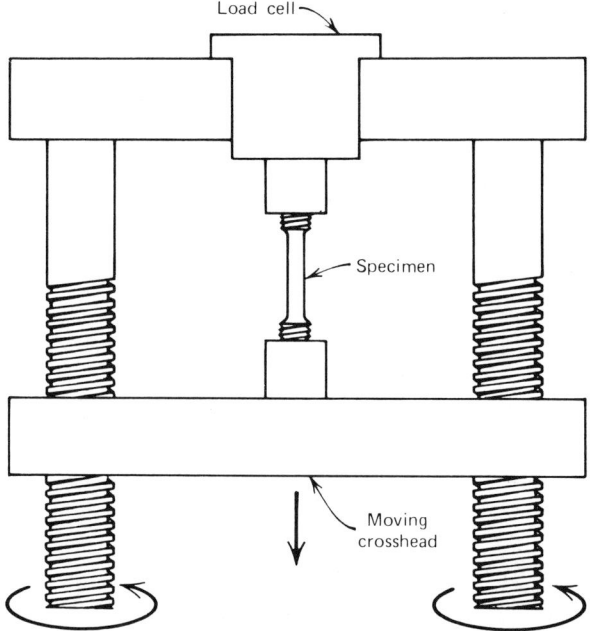

Fig. 2-23. Tensile test machine. The two screws drive the horizontal bars apart, stretching the specimen. The load cell measures stress. The strain can be measured with a mechanical gauge. (From Hayden HW, Moffatt WG, Wulff J: The Structure and Properties of Materials. Vol 3: Mechanical Behavior. John Wiley & Sons, New York, 1965, with permission.)

Stainless Steel

Stainless steels have been used for implants for about 60 years. The best stainless alloy for surgical use is type 316L, which has 17% to 20% chromium, 10% to 14% nickel, 2% to 4% molybdenum, very low (< 0.08%) carbon, and the rest iron. These alloys can be forged, and they harden (ie, the yield strength increases) as the amount of cold forging or plastic deformation increases, but at the expense of some ductility.* They can be made softer (weaker) and more ductile by *annealing* (heating in a suitable furnace). Thus, in severely forged 316L stainless steel the yield stress may be as high as 875 million newtons/m^2 (125,000 psi) with an elongation of 15% (or less)

---

* *Forging* is essentially, in this particular context, the art of the blacksmith. The metal is heated and hammered or squeezed into shape. This may or may not be done with aid of a die, which is a mold to guide the flow of the metal. Alternatively, metal may be shaped by *casting* (ie, melting and pouring into a mold) and subsequent solidification.

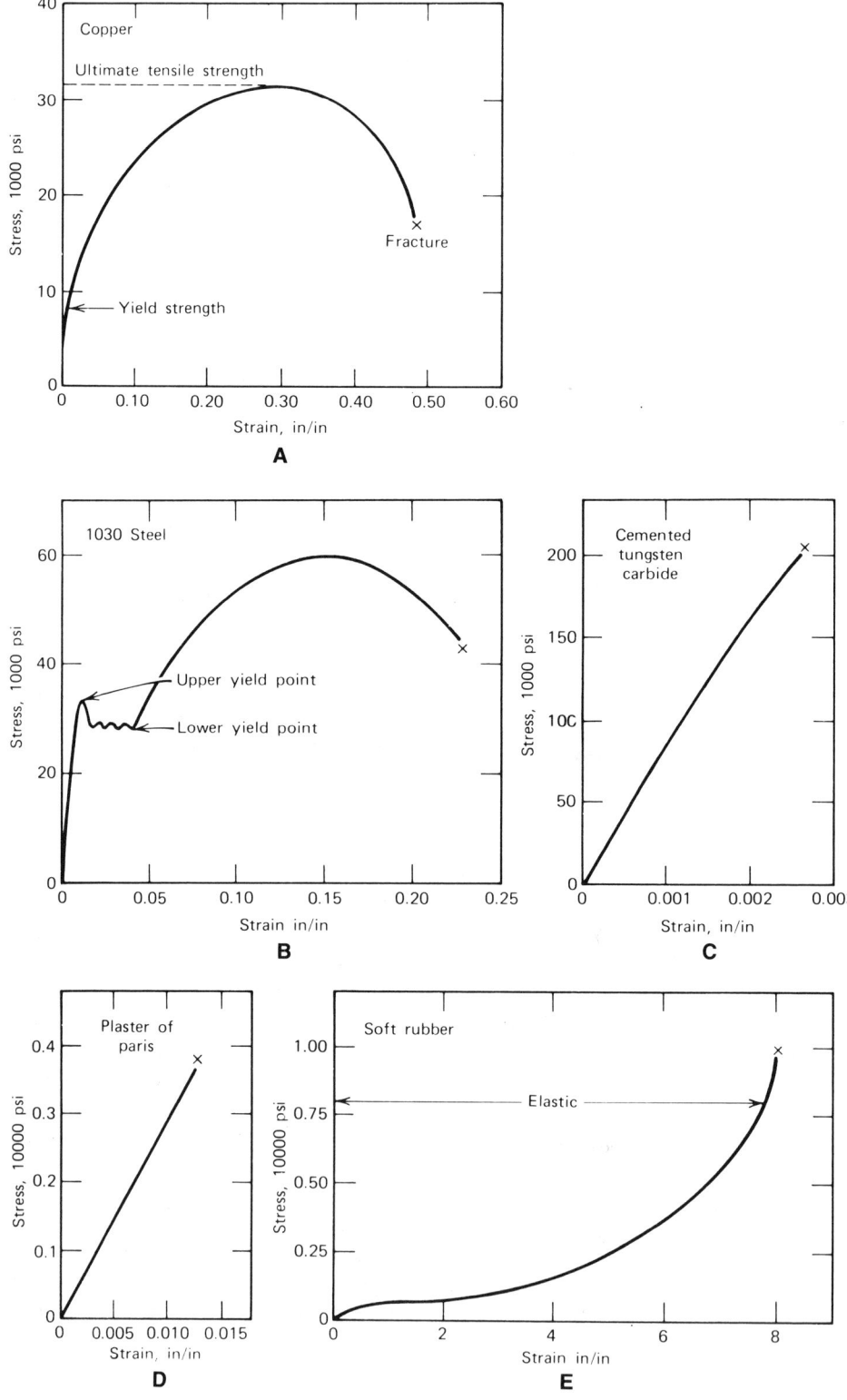

at fracture, whereas a thoroughly annealed piece of the same material may have a yield stress of 210 million newtons/m^2 (30,000 psi or less!) but an elongation of 50% or more.

For load-bearing implants, forged material is generally required. Cast stainless steel is not as strong as forged stainless; yield stresses of 210 million newtons/m^2 or less are common for the cast material. In addition, the castings are chemically inhomogeneous and therefore have inferior corrosion resistance as well. Thus cast stainless steel is poorly suited to implant applications. Forged stainless steel has superior mechanical properties and is relatively low in cost. However, even the best 316L has only marginal in vivo corrosion resistance. It has been shown that in typical clinical applications of multicomponent implants (eg, plate with screws) some corrosion is inevitable in this material. Furthermore, corrosion resistance decreases as strength increases; thus there is a compromise between mechanical properties and corrosion resistance.

Stainless steel is the material of choice for fixation devices that can be removed in a year or two.

Cast Cobalt-Chromium-Molybdenum Alloy

Discovered 80 years ago, this alloy was used for tools and similar applications. Under the trade name of "Stellite 21," it was soon used for dental restorations and, about 50 years ago, for orthopedic implants, mainly fixation devices. The alloy's composition is 63% to 70% cobalt, 25% to 30% chromium, and 5% to 7% molybdenum, with minor amounts of impurities such as manganese, nickel, iron, and carbon. It is sold under various trade names.

This alloy is much more resistant to in vivo corrosion than is 316L stainless steel (or any other stainless steel). The chief disadvantage of cast Co-Cr-Mo alloy is that its mechanical properties are inferior to those of stainless steel (see Table 2-1). Usually, the alloy is cast by a relatively expensive process that generates a variety of microscopic defects that limit its strength, ductility, and fatigue life. It is the material of choice for permanent implants because of its superior in vivo corrosion resistance and also because such implants can usually be more massive than fixation devices. It is also chosen for applications where the implant shape must be cast, because of complexity, rather than forged.

Fig. 2-24. (A) Tensile test data for copper as usually plotted. Engineering stress-strain curves for several engineering materials. **(B)** 1030 steel. **(C)** Cemented tungsten carbide. **(D)** Plaster of Paris. **(E)** Soft rubber. (From Hayden HW, Moffatt WG, Wulff J: The Structure and Properties of Materials. Vol 3: Mechanical Behavior. John Wiley & Sons, New York, 1965, with permission.)

Table 2-1. Mechanical Properties of Implant Alloys

	316L Stainless Steel	Cast Co-Cr-Mo Alloy	Advanced[a] Co-Cr-Mo Alloys	Titanium "Six-Four" Alloy
Yield stress (N/m^2)	250–900 × 10^6	500 × 10^6	800–1200 × 10^6	815 × 10^6
(psi)	35–125,000	70,000	110,000–160,000	145,000
Ultimate tensile stress (N/m^2)	550–1000 × 10^6	700 × 10^6	970–1900	1200 × 10^6
(psi)	80–145,000	100,000	140,000–280,000	165,000
Elongation at fracture (%)	45–15	8.0	5.0–15	15

[a] These figures may change as the new technologies develop. See text.

Other Wrought Cobalt-Base Alloys

A sister alloy, "Stellite 25," discovered at about the same time as the cast Co-Cr-Mo alloy, is much more ductile and can be forged. This alloy is 52% to 58% cobalt, 19% to 21% chromium, 14% to 16% tungsten, 9% to 11% nickel, and miscellaneous minor impurities. Its mechanical properties are superior to those of 316L stainless, but inferior to the cast Co-Cr-Mo alloy discussed above. It is important not to confuse the wrought Co-Cr-W-Ni alloy with the cast Co-Cr-Mo alloy, as it is not good practice to combine them in multicomponent prostheses. For instance, nail-plate assemblies with plates of the wrought alloy and nails or screws of the cast alloy have been shown to corrode in vivo and cause clinical problems.

Another alloy, "MP35N," is nominally 20% cobalt, 10% molybdenum, 35% nickel, and 35% cobalt. Again, the corrosion resistance is better than that of stainless steel, but inferior to the cast Co-Cr-Mo alloy. Very high strengths can be achieved with this material (eg, tensile strengths in excess of 1,400 million newtons/m^2; over 200,000 psi). In general, it is preferable to use the advanced versions (see below) of the Stellite 21 composition in order to achieve higher corrosion resistance as well as high strength.

Advanced Cobalt-Chromium-Molybdenum Implant Alloys

The possibility of improving the mechanical properties of the cast cobalt-chromium-molybdenum (Stellite 21) alloy has been the subject of considerable attention, since the in vivo corrosion resistance is excellent, uniquely so. Wrought material, hot isostatically pressed cast material (in essence, castings that have been subjected to high pressures at very high temperatures), and hot isostatically pressed powder processed material (ie, material fabricated by subjecting alloy metal powders to hot isostatic pressing) of the above-mentioned composition have been available on a limited basis, usually in femoral components of total hip replacements. Extremely high strengths (eg, over 1,650 million newtons/m² or 240,000 psi) can be attained by these advanced technologies, which are limited by cost considerations and also by shapes that can be fabricated. Some of the wrought material has been reported to have superior corrosion resistance to even the cast alloy (see Devine TM, Wulff J, J Biomed Mater Res 9:151, 1975).

Titanium

Commercially pure titanium is highly resistant to corrosion but has low yield and ultimate tensile stresses. Therefore it is not considered generally suitable for fracture fixation devices.

Titanium "Six-Four" Alloy

Titanium "six-four" alloy has 5.5% to 6.5% aluminum and 3.5% to 4.5% vanadium in a titanium base, with few impurities permitted. The usual grade offered for implant fabrication is the ELI grade, with specially low levels of carbon, oxygen, nitrogen, and hydrogen; not uncommonly, the sum of these impurities is below 0.1% in such a grade. Such purity enhances ductility and resistance to fracture even in the presence of stress concentrations. This alloy, when forged and properly heat treated, has excellent mechanical properties that are superior to all of the above-mentioned materials, and extreme resistance to crevice corrosion as well. (Like stainless steel, the cast version of this alloy is markedly inferior to the wrought version in mechanical properties.) Corrosion products of titanium appear to be less inflammatory than other metals. Since titanium alloys are recent arrivals compared with the other implant materials, some questions remain; however, the general clinical experience for the two decades these alloys have been implanted has been very good. The one concern with regard to titanium alloys has been wear resistance, even in contact with polyethylene. At this time, the wear rate of articular joint prostheses using titanium and titanium alloys is unacceptably high. However, current research using ion implantation and other surface treatments may solve this problem.

Interaction of Implant Materials With Bone: Osteopenia and Attachment

Table 2-1 compares the mechanical properties of the alloys discussed in this section. For the reasons discussed above, the most popular alternatives are cast Co-Cr-Mo and 316L stainless steel. The latter offers superior mechanical properties, low cost, and reliability (freedom from casting defects), but

Table 2-2. Young's Modulus of Bone and Typical Engineering Materials

	N/m²	psi
Tungsten	395.5×10^9	56.5×10^6
Stainless Steels	$196-210 \times 10^9$	$28-30 \times 10^6$
Titanium Alloys	$105-119 \times 10^9$	$15-17 \times 10^6$
Bone: Cortical	$7-21 \times 10^9$	$1-3 \times 10^6$
Cancellous	$0.7-4.9 \times 10^9$	$0.1-0.7 \times 10^6$
Polymethyl Methacrylate	$2.5-3.5 \times 10^9$	$0.35-0.5 \times 10^6$
Polyethylene	$0.14-0.42 \times 10^9$	$0.02-0.06 \times 10^6$

at the expense of corrosion resistance. A recent study of multicomponent stainless steel implants removed for various clinical reasons showed that more than 90% had corroded. Thus stainless steel should be used in fracture fixation devices that cannot be too massive and rigid, therefore bearing higher stresses, and that can be removed after the fracture heals. For prosthetic implants or any device that must remain in place indefinitely, the Co-Cr-Mo alloy, whether cast, wrought, or fabricated by powdered technology methods, is preferable.

As Table 2-2 shows, high-strength engineering materials are much stiffer than bone. Thus a fixation device made massive to avoid fatigue failure is very stiff, bearing much more of the normal load that would be taken up by the bone. Normal healing and remodeling of the bone would therefore be impeded. Thus proper fixation devices are limited in strength and should be regarded as essentially alignment aids. They should not be expected to bear full ambulatory loads for extended time periods. If nonunion occurs, fatigue failure of the device is inevitable. A clinical advantage of high-strength materials for fracture fixation is that the fatigue life is longer (at equivalent loads), and slowly healing fractures may be treated without fear of interruption by implant failure.

It is appropriate—especially in patients with normal bone—to remove the fracture fixation device as soon as healing is adequate. This allows daily-activity-generated stresses to pass through the bone and stimulate remodeling. As long as the implant is present, much of the load is borne by the stiffer metal and the bone under the metal becomes osteopenic. Refracture is therefore most likely in the bone immediately adjacent to the end of a plate, as a result of stress concentration and the osteopenic bone under the plate, as shown in Figure 2-25.

Another interaction of implant materials with bone is ingrowth into any porosity that exceeds a certain size (typically of the order of 10^{-4} m, depending on the specific material). This tendency has been exploited in order to produce cementless joint replacements that attach themselves by ingrowth of the bone in contact with the prosthesis. Figure 2-26 shows the surface of one such prosthesis. The surface has been produced by heating spherical metal powders of uniform size at high temperature in contact with

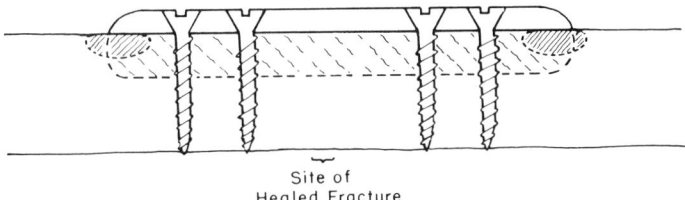

Site of Healed Fracture

Fig. 2-25. The plate lowers the stress in the adjacent bone (lightly shaded areas), leading to osteopenia. Because of the discontinuity in stiffness there is a stress concentration at each end of the osteopenic region.

the body of the prosthesis; the result is a layer of porous metal. This technique can only be used with extremely corrosion-resistant materials, either the cast cobalt-chromium-molybdenum alloy or titanium alloys, because of the many crevices generated. The high temperatures necessary to produce the porous layer compromise the mechanical properties of the substrate.

Polymers, Plain and Fiber-Reinforced

Ultra-high-molecular-weight polyethylene is universally accepted as a bearing material in articular prostheses because of its low wear rate in contact with cobalt-chromium alloys and stainless steels (wear rates in contact with titanium alloys are unacceptably high). However, ultra-high-molecular-weight polyethylene is not strong enough to serve as a structural material as its tensile strength is usually less than 5,000 psi (about 35 ×

Fig. 2-26. The surface of a "cementless" femoral component. The spheres are approximately 10^{-4} in diameter.

Table 2-3. Typical Tensile Strength and Young's Modulus of Polymers, With and Without Fiber Reinforcement

	Tensile Strength		Young's Modulus	
	N/m²	psi	N/m²	psi
Polyethylene	35×10^6	5×10^3	0.3×10^9	0.05×10^6
Nylon	80×10^6	12×10^3	2.8×10^9	0.4×10^6
Polyesters	40×10^6	6×10^3	5.5×10^9	0.8×10^6
Polyphenylene Sulfide	140×10^6	20×10^3	12×10^9	1.7×10^6
Polyester, glass, fiber reinforced, 50%	160×10^6	23×10^3	16×10^9	2.3×10^6
Kevlar-Epoxy, 54%	1200×10^6	172×10^3	83×10^9	12×10^6
Graphite-Epoxy, 60%	1500×10^6	220×10^3	110×10^9	16×10^6

10^6 newtons/m²). Table 2-3 shows that unreinforced plastics have relatively low tensile strengths and elastic moduli. These properties improve dramatically when fiber reinforcements are incorporated into the plastics. Considerations of toxicity and carcinogenicity limit the choice of fiber (and polymer matrix) for implant applications. Graphite fibers are relatively innocuous in these respects and have been used (with varying degrees of success) to reinforce acrylic cement and polyethylene for surgical applications. Epoxy compounds, such as those included in Table 2-3, are not. One advantage of the polymers (eg, for fracture fixation) is their low elastic moduli, which would limit the contact osteopenia described in Figure 2-25 if plastic fixation devices were used. Such applications are limited by the low strength of the plastics. When the strength is increased by fiber reinforcement, the elastic modulus also increases (Table 2-3), so that the advantage tends to vanish. Although polymers do not corrode, they do deteriorate over time, usually because of oxidation. There is evidence that this occurs in femoral and tibial components of total hip and knee joint prostheses.

8 MECHANICAL CONSIDERATIONS IN TREATMENT OF FRACTURES

Before considering internal fixation devices on fractured bones, let us consider mechanical factors involved in the functional treatment of fractures. Joint motion maintained through the period of fracture healing obviates the long-term physical therapy frequently necessary to regain motion in a joint whose capsule has been immobilized for long periods of time and is scarred down. Also, intermittent loading speeds fracture healing. Whether this occurs by generating the electrical potentials that have been seen in bone subjected to intermittent loading or bending or is merely an incidental reflection of mechanical stress is not certain. Generation of electrical activity is a property common to almost all organic material and is a result of the fact that organic molecules may be asymmetric or arranged so as to have an asymmetric charge distribution.

Consider the two arrangements in Figure 2-27. The positive and negative charges are symbolized by full and open circles, respectively. In Figure 2-27B, the arrangement has no symmetry across the horizontal plane, and a voltage results as the material is compressed. Figure 2-27A is symmetric across the plane of compression, and no voltage appears. Quartz crystals are essentially equivalent to the arrangement in Figure 2-27B and are *piezoelectric*; rock salt, which is equivalent to the arrangement in Figure 2-27A, is not. Whether this is meaningful physiologically or is just a coincidental physical happening is not clear, but there is no question that the application of intermittent load—whether it be electrical or physical—does speed fracture healing.

The relationship between metabolic activity and mechanical stress has been known for years. The older German literature suggests that mesodermal (connective tissue) primitive cells subjected to pure tension and pure compression tend to form bone, as shown schematically in Figure 2-28. Similar cells subjected to shear form fibrous tissue and such cells, under equal

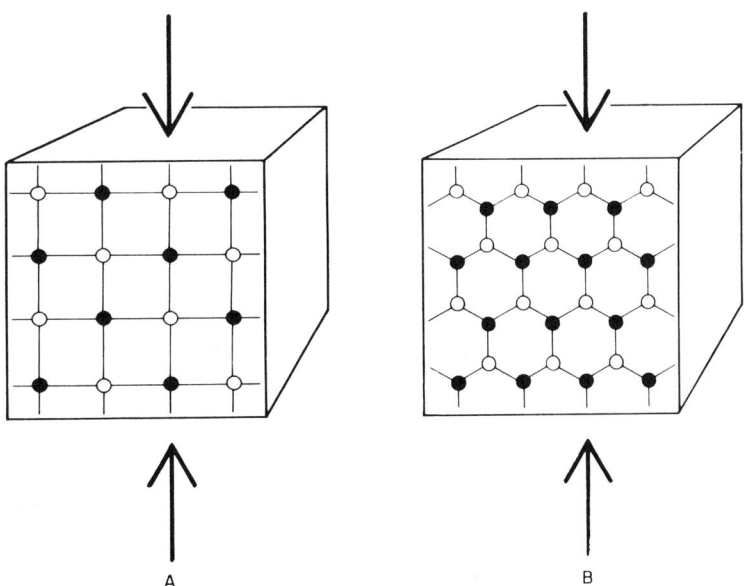

Fig. 2-27. (A) Schematic "rock salt" crystal: full circles represent Na^+; open circles are Cl^-. Across any horizontal plane there is symmetry; a sodium ion on one side is in a position identical to a sodium ion on the other. Thus, when the stress shifts the ions, all charge movements cancel and no voltage is induced. **(B)** Crystal with no symmetry across the horizontal plane. A voltage is produced (ie, it is piezoelectric).

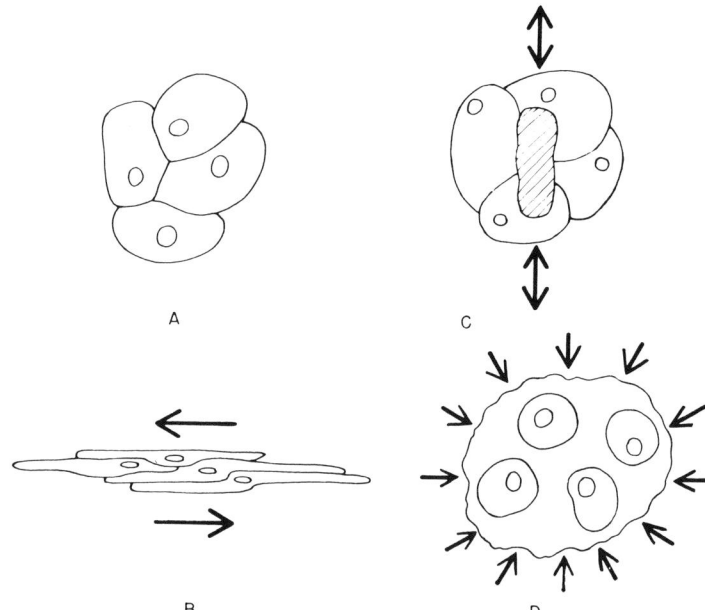

Fig. 2-28. Metaplasia of fibroblasts into various types of connective tissue under the influence of mechanical stress has been reported. Primitive cells **(A)** under shear create fibrous tissue **(B)**; under pure tension or compression, bone is formed **(C)**, and hydrostatic pressure results in a cartilaginous matrix **(D)**.

pressure (hydrostatic pressure) from all directions, form cartilage. (The implications of this in the formation of reparative articular surfaces are discussed in Chapter 4.) Thus stress primarily in pure tension or pure compression and intermittent in nature aids in fracture healing. More recent observations indicate that certain amounts of motion are compatible with healing fractures and that some shear is permissible if it is small. Apparently, limited shear aggravates the formation of fracture callus, and more callus is laid down around fractures that have slight motion than around those that are rigidly immobilized. Certainly significant shear or bending at fracture sites is clinically known to lead to nonunion and, if the motion is great enough, to formation of a pseudoarthrosis.

The most effective way to transmit a stress, which is limited in shear and bending (and, of course, torsion) but is mainly of tensile or compressive nature, is with a cast brace or weight-bearing cast. Tibial fractures submitted to weight-bearing in a cast brace heal two to three times faster than similar fractures treated in nonweight-bearing plasters or with the patient on crutches. Such cast-brace treatment is associated with proliferative callus, which acts as a scaffold, immobilizing the fracture fragments. This method of fracture treatment is in marked contradistinction to rigid internal fixation where external callus formation is minimal and sometimes even nonexistent. Here again the basic principle is to eliminate bending and torsional stresses on the fracture site as much as possible.

9 INTERNAL FIXATION DEVICES: WIRE AND TENSION BANDS

The role of mechanical stress in fracture healing is not completely understood and is in some respects controversial. However, satisfactory results are known to occur when fracture fragments are in reasonable apposition and relative motion of the fragments is limited.

In general, muscular activity tends to bring fracture fragments together. In certain places on the skeleton, individual muscles can act to separate fragments. The iliopsoas tendon, for example, pulls the entire lower extremity below the femoral neck fracture into external rotation. In general, however, tensile forces result physiologically from bending and torsional stress and are also due to muscle forces. The purpose of fracture fixation devices is therefore alignment and resistance to tensile stress. Hence, all fracture fixation devices can be considered as bands that resist tension and are most effective if placed on the tensile side of the fracture.

The simplest demonstration of this general principle in internal fixation is wire fixation. Consider the wiring of a patella fracture (Fig. 2-29). The wire ties together opposing points on the anterior cortex about the point of contact. A *torque* equilibrium is maintained for the distal fragment (Fig. 2-29), for a small angle of flexion. The moment of the tendon force is balanced by the reaction force from the other fragment (note that the reaction force is compressive). In other words, the wire is attached so that the tendon force

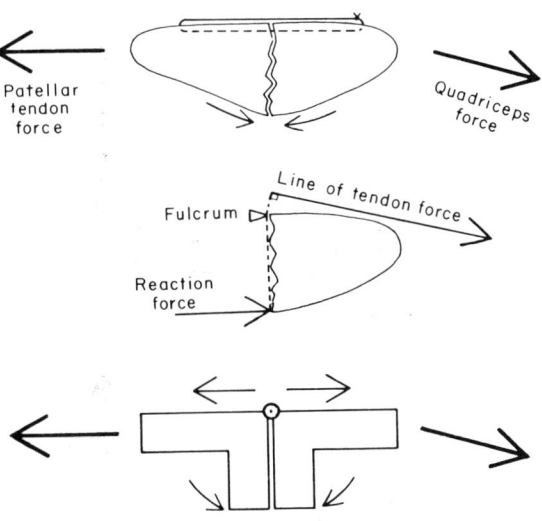

Fig. 2-29. The quadriceps-patellar tendon forces produce compression in the fragments when the anterior cortex is wired.

86 PRACTICAL BIOMECHANICS FOR THE ORTHOPEDIC SURGEON

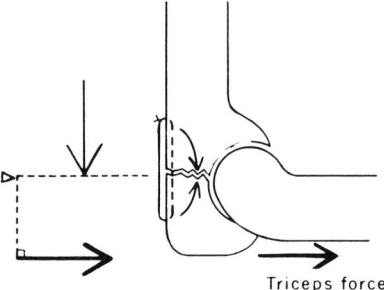

Fig. 2-30. Correct wiring of the olecranon converts muscle forces into compression.

rotates the distal fragment into contact with the proximal fragment, and larger tendon forces cause larger compressive forces across the fracture surface. All this occurs provided that contact at the fulcrum (the anterior cortex) is maintained by tension in the wire. The reaction force and the tendon force both have appreciable components in the same direction, which can only be balanced by a tensile force in the wire, thus the phrase *tension band*. An analogous situation with two pieces of hardware connected by a hinge is also shown in Figure 2-29.

Another example is wire fixation of the olecranon process of the elbow (Fig. 2-30). The wire anchors the two fragments together, as close to the outer surface as possible, so that the triceps force rotates the two fragments together and keeps them together with a compressive reaction force, which is required to maintain torque equilibrium about the "hinge." The point of contact is maintained by the wire. By considering full force equilibrium in

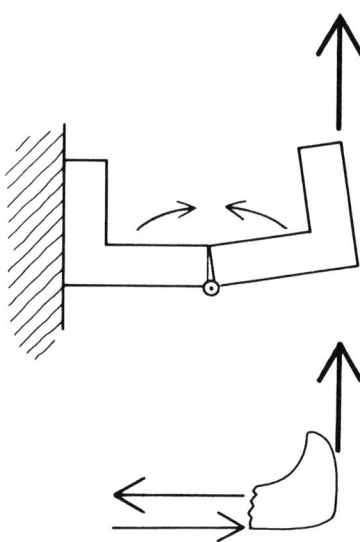

Fig. 2-31. A mechanical analogy to the forces on the olecranon in Figure 2-30.

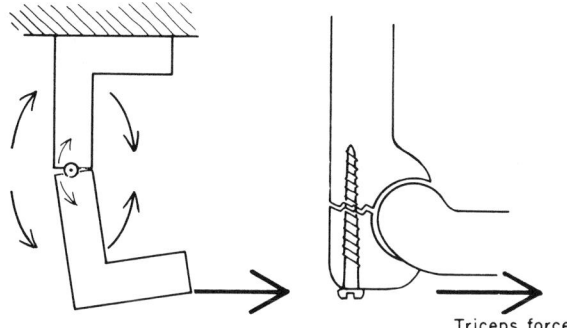

Fig. 2-32. Screw fixation of the olecranon moves the pivot to the interior so that only one side of the fracture is compressed by the muscle force.

the horizontal direction, we can see that the compressive reaction force must be balanced by an equal and opposite force in the wire (Fig. 2-31).

Properly done, the wire method is much more effective than using a screw. As shown in Figure 2-32, with screw fixation the "hinge" or center of rotation is no longer at the central portion of the fracture. Thus, the outer portion of the fracture, as shown by the "hardware" analogy, is free to open up when the triceps force is applied; this tendency is counteracted only partially by the rigidity of the screw. In addition, the screw is subjected to bending stresses. Screws are not designed (or intended) to withstand significant bending stresses, and the latter should be avoided or minimized.

10 INTERNAL FIXATION DEVICES: PLATES

A multiple-screw plate, mounted on the extreme fiber in tension of a long bone, is somewhat similar to the wire fixation cases discussed above. As Figure 2-33 shows, if the plate is properly located, bending moments lead

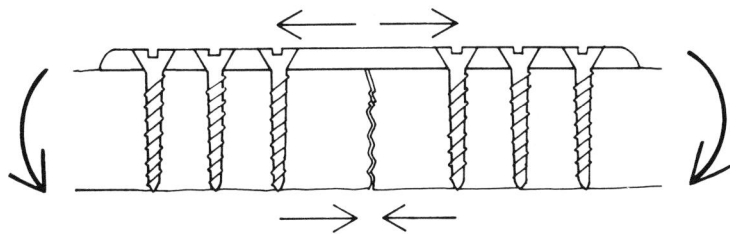

Fig. 2-33. When the plate is located at the tensile surface, the fracture is compressed by the muscle forces across most or all of the fracture surface.

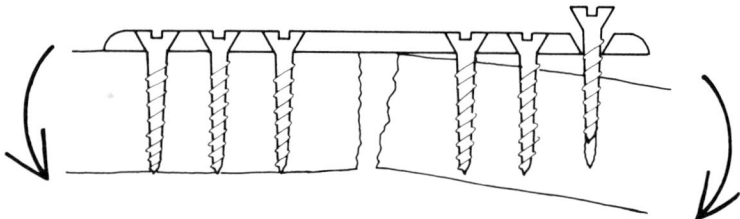

Fig. 2-34. If the plate is not tightly fastened to the bone, bending stresses occur on the screws.

to compressive forces across the fracture. In this case, although the plate participates in bending because it is at the extreme fiber, the stress in the plate is essentially tensile in nature (as in the case of wire fixation) for the ideal case of perfect apposition. The screws serve to constrain the bone-plate composite system so that they act together in bending, with the plate in tension.

There are a number of complications and potential sources of difficulty in this arrangement. For instance, if the plate is not tightly set against the bone by the screws, the screws may have bending stresses on them (Fig. 2-34). At least two screws are necessary on each side of the fracture to prevent rotation caused by moments that can occur at right angles to the principal bending moment. Since apposition is rarely (if ever) perfect, the plate is usually subjected to a certain amount of bending with little or no assistance from the bone. Thus, if the plate is too thin, it fails in fatigue because of repeated bending. If a screw is omitted from a hole near the fracture, the plate may bend excessively at the empty screw hole. It is not constrained to the bone, and the screw hole is a weak point. This stress concentration causes fatigue failure at the screw hole.

One way to increase the supporting function of the bone and to decrease the bending of the plate is to use a compression plate. This plate is designed to put the fracture in compression, even in the absence of forces or move-

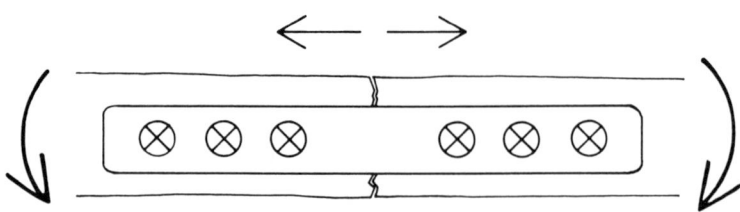

Fig. 2-35. Arrangement 10 to 100 times as rigid as that in Figure 2-34 and superior to that in Figure 2-33 when apposition is poor.

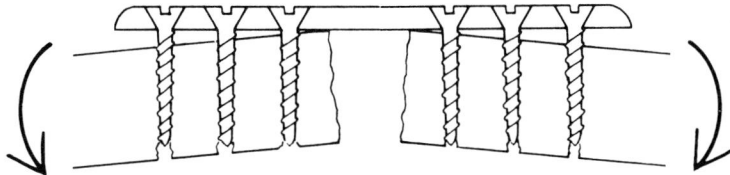

Fig. 2-36. Likely consequence of single plate fixation with a gap at the fracture site.

ments from the rest of the musculoskeletal system. Various mechanical devices and modifications are available to achieve this effect. Specially shaped screw holes force the screw (and therefore the bone) to move along the plate toward the fracture side as the screw is tightened. Other devices force the bone together in compression before the screw holes are drilled and the plate installed.

Each method has unique advantages and disadvantages, but all methods put the bone in compression and the plate in tension. Since the fracture surface is irregular, the compressive load is borne by a relatively small portion of the fracture surface (ie, by asperities in contact). After about 72 hours these asperities are resorbed and the compressive stress in the bone (and tensile stress in the plate), because of the compression device, has generally vanished. The apposition of the fracture is greatly improved, and the ideal case shown in Figure 2-33 is approached. With poor apposition the plate alone bears the bending moment (Fig. 2-34). In this situation the plate alone has little rigidity compared with the bone-plate composite system (Fig. 2-33). Typically, the plate alone has less than 1% of the rigidity of the bone-plate system acting together (Fig. 2-34).

Of course, the plate may be mounted at right angles to those in Figures 2-33 and 2-34 in the plane of bending (Fig. 2-35). This arrangement works well if apposition is poor. The rigidity is much greater—10 to 100 times as great as for the situation in Figure 2-34. However, the screws are now subjected to bending and torsional forces, and these rather than the plate may fail. When apposition is good, this fixation is inferior because the plate is centered at the neutral axis in bending rather than at the extreme fiber, so that total rigidity is less than for the case of Figure 2-33, and the fracture is not totally in compression. Single plate fixation is thus not recommended when a gap exists at the fracture site. No matter where it is relative to the bending axis, the plate will probably bend or the screws loosen (Fig. 2-36). A comminuted fracture can create the same problem.

Two methods of using two plates to cope with "gaps" are shown in Figure 2-37. Such fixation is most effective when the axis of the applied moment(s) is well known. In Figure 2-37B, the two plates suffice to make the fixation act as a composite beam, with one plate in tension and the other in compression, with good rigidity. However, even if not on opposite sides of the bone, two plates provide significantly increased torsional rigidity.

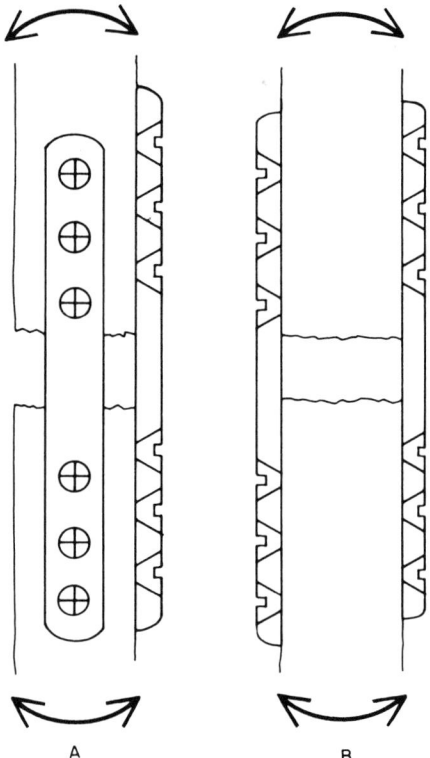

Fig. 2-37. Two-plate fixation of gapped or comminuted fractures.

11 SPIRAL FRACTURES

Section 3 describes how shear stress can result in tensile stresses, which are largest on a plane 45° from the plane of shear. Torsional fractures occur at 45° to the axis of torsion, and the fracture crack, to remain at 45° to the applied shear stress, must follow a helical path (ie, it is a "spiral" fracture). In the fixation of such a fracture one must consider this. The forces that tend to disrupt it—the same that tend to disrupt any fracture—are bending and torsion.

If plates are used across the fracture surface but at a 45° angle to the axis of the bone (Fig. 2-38A), the plates resist torsion (on the bone) well, but bending not so well (Fig. 2-38B). In particular, the "tension band" is forced to act at a 45° angle to the extreme fiber stress, which it does not do well. This does not happen if the plate is parallel to the bone; the problems resemble those in Figure 2-34 in bending! A considerable moment is developed about the screws (Fig. 2-39) when torsion is applied.

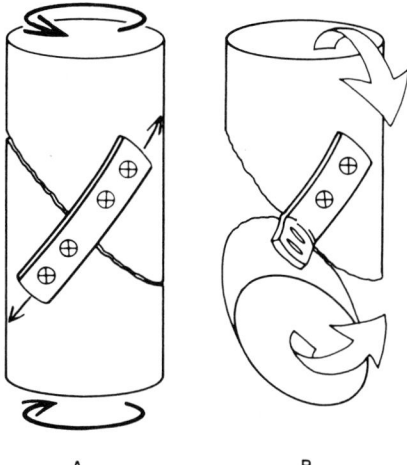

Fig. 2-38. Plating across the fracture surface of a spiral fracture resists torsion well **(A)** but bends poorly **(B)**.

A mechanically attractive solution to this problem is to use straps or wires (Fig. 2-40) drawn so tightly that rotation of the fragments becomes impossible. This solution, however, is not usually biologically attractive. It can lead to circulatory problems, and the high stresses under the wire can cause sufficient resorption to loosen the wires to the extent that there is no longer resistance to torsion.

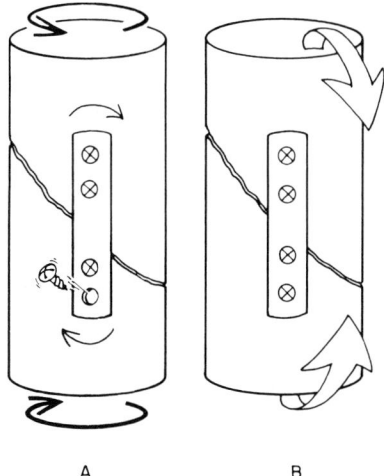

Fig. 2-39. Plating a spiral fracture along the bone axis resists bending reasonably well **(B)** but torsion poorly **(A)**.

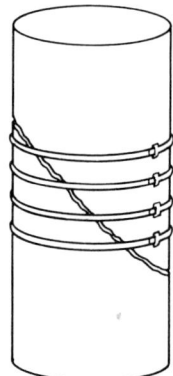

Fig. 2-40. Biologically objectionable way to cope with a spiral fracture.

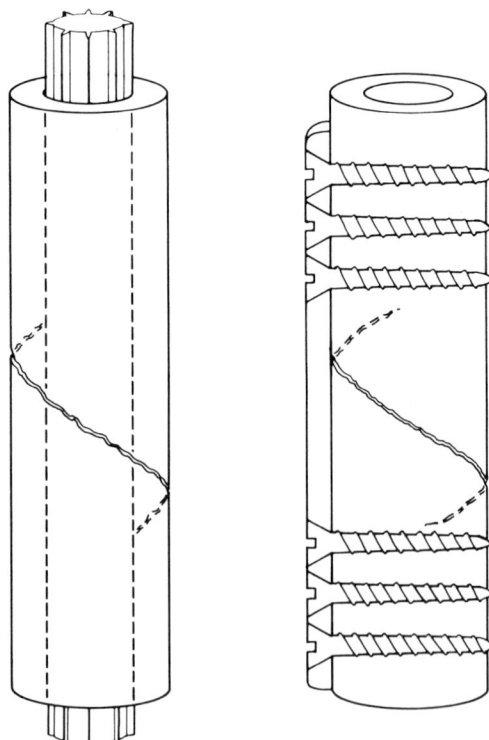

Fig. 2-41. Spiral fracture fixed with a fluted intramedullary rod (*left*) or a long plate (*right*).

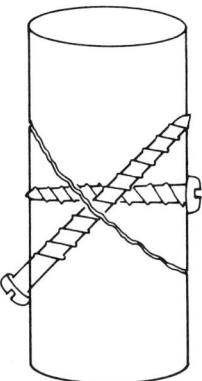

Fig. 2-42. Screw fixation of spiral fracture; screws placed to optimize bending and torsional stability.

Two viable fixation methods for noncomminuted spiral fractures are shown in Figure 2-41: a fluted intramedullary rod and a long plate. The flutes in the rod are necessary for torsional stability. This can also be achieved by locking an unfluted intramedullary rod (see Fig. 2-58, below). The length of the plate serves to reduce the forces corresponding to an applied torque. The moment arms of the forces on the screws are greater, so less force is needed to balance torques and bending moments.

When plates or rods are undesirable, spiral fractures may be fixed with screws, but with the same problem again. If the screw is placed perpendicular to the fracture surface, optimal torsional stability is obtained, as the screw is parallel to the tensile stress. However, in such a situation, bending moments on the bone cause bending moments on the screw. For optimal bending rigidity the screw should be perpendicular to the axis of the bone rather than the fracture surface. Where both twisting and bending moments are anticipated (e.g., the tibia), one screw may be placed in each orientation (Fig. 2-42). Now one screw relieves the bending stresses on the other, whether the bone is in torsion or bending.

12 SCREWS

The "tension band" framework includes screws. Screws are used to compress fracture fragments together or to hold a plate against bone by compressive force. The balancing force, as in the case of wires and plates, is a tensile force in the fixation device. The screw creates this tensile force by being an elementary machine that is commonly used to convert a small torque to a large axial force. This large mechanical advantage, which makes screws generally useful as fasteners where large holding forces are desired, is par-

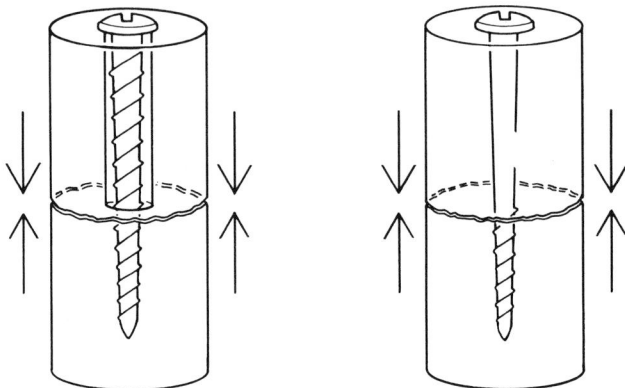

Fig. 2-43. Lagging the screw and use of a "lag screw" to ensure compression across the fracture surfaces.

ticularly useful for the fixation of small fracture fragments. To maximize the holding force, the screw must be "lagged" (Fig. 2-43) so that its threads do not separate the fragments.

A *"self-tapping" machine screw* with cutters at its tip may be inserted in the drilled hole, or a tool (a "tap") may be used to cut threads in the hole and a machine screw with a smooth end used. To minimize local fracturing and mechanical damage near the screw in bone, the screw hole should be drilled and tapped (ie, threads must be cut in the hole). The cutting edges of the tap are necessarily sharper than the tip of a self-tapping screw, because the tap may be more carefully (and expensively) made, and tap materials may be high hardness tool grade alloys that cannot be left in the body because of inadequate corrosion resistance. A sharper cutting edge involves smaller forces and necessarily leaves fewer fragments. A tap cuts cleaner threads than a self-tapping screw and also permits the hole to be flushed of debris before the screw is inserted.

If a single helical thread is cut into a screw, then the pitch (defined as the distance along the screw axis between two parallel loops of the helix) is equal to the lead (the distance the screw advances with each turn), as shown in Figure 2-44.* Since pullout of a screw is accomplished by shearing the bone, the holding strength of a screw is roughly proportional to the area of a cylinder with length equal to the screw length and diameter equal to the outer diameter of the screw (Fig. 2-45A).

---

* A screw (or hole!) can be made with two or more helices, running in parallel (ie, it can have multiple threads). Multiple-threaded screws advance faster as they are turned: the lead of a double-threaded screw is twice the pitch, so it advances twice as fast as a single-threaded screw having the same pitch.

12. SCREWS

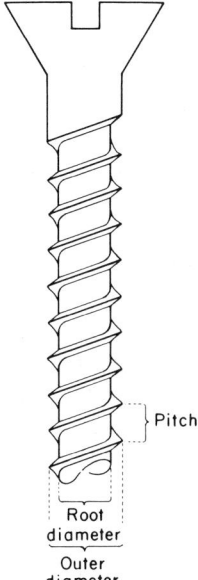

Fig. 2-44. Descriptive terms for screw threads.

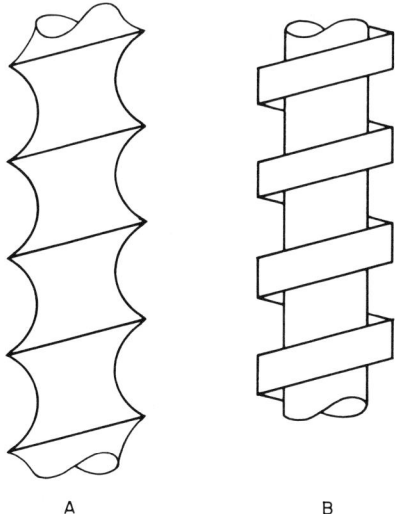

Fig. 2-45. (A) Thread design with high holding power and fatigue resistance. **(B)** Thread design with low holding power.

If the thread shape is bulky or squared (Fig. 2-45B), so that bone is removed from this area by the screw, then holding power is reduced. The thread shape in Figure 2-45A is much better in this respect. Screws with sharp threads can be used only in pre-tapped holes. Also notice that the thread roots are rounded to avoid stress concentrations in bending or tension of the screw. This practice greatly enhances the fatigue life of the screw.

13 NAILS, RODS, AND PINS

Nails, rods, and pins, in contrast to plates and screws, are intentionally subjected to bending and/or torsion. Consider the femoral neck fracture shown in Figure 2-46. Obviously the force on the femoral head will have a bending moment at the fracture. How large this moment can be is shown in Figure 2-47, where the force and moment equilibriam between abductor muscle force and body weight are considered for the left hip when the right foot is off the ground (eg, while walking). We can, in our minds, isolate the rest of the body from the left leg and consider the forces transmitted to the rest of the body from that leg. These forces consist mainly of the abductor muscle force and the hip joint force. The hip joint is like a fulcrum: since the right foot is not on the ground, the abductor muscle acts to keep the

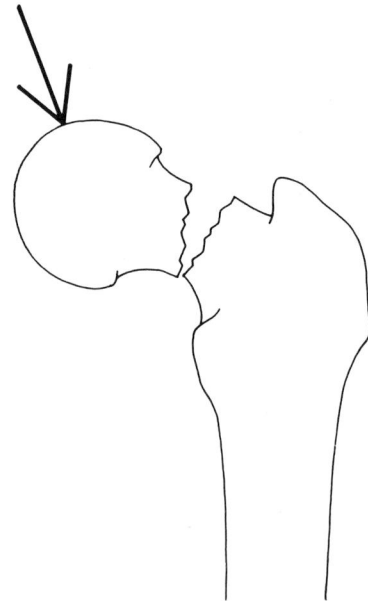

Fig. 2-46. The ambulatory force on the femoral head has a bending moment at the fracture site.

13. NAILS, RODS, AND PINS

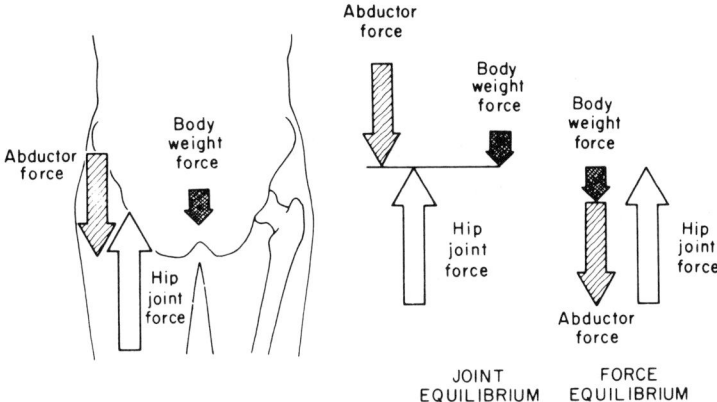

Fig. 2-47. Because of its small lever arm about the hip joint, the abductor muscle force must be large to achieve torque equilibrium. The sum of the abductor force and body weight must equal the force on the hip joint to achieve force equilibrium. Thus the force across the hip joint is large, typically four times body weight.

trunk level and prevent rotation to the right by supplying a moment to balance the body weight on the other end of the "lever." However, the lever arm of the abductor muscle is relatively short. Typically it is one third of the lever arm of the body weight, with the hip joint as fulcrum. Thus the abductor muscle force may be three times body weight.

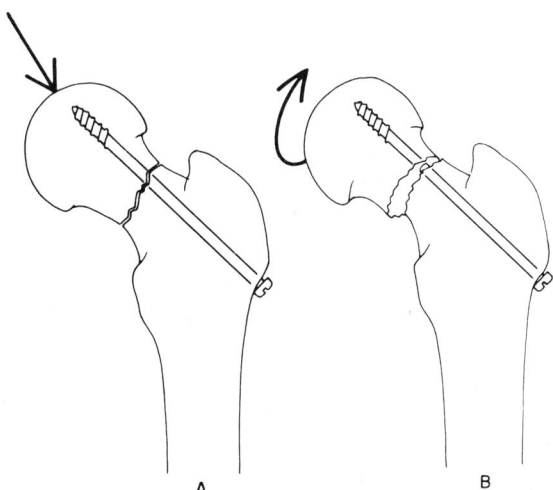

Fig. 2-48. Placement of the pin as shown **(A)** resists the bending moment of the normal joint load (see Fig. 2-47) but **(B)** does not resist any twisting motions.

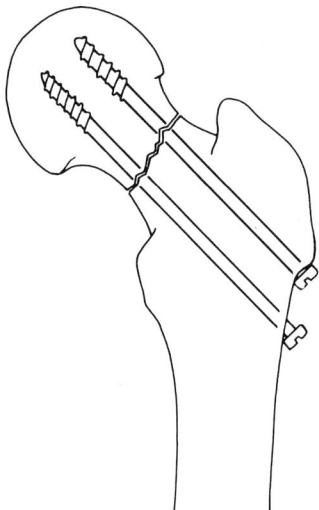

Fig. 2-49. It requires at least two pins to resist twisting of fractured femoral necks. Pin placement in such circumstances depends on both the mechanical considerations and the resistance of pin pull-out.

Since force equilibrium requires that the hip joint force equal the sum of the muscle force and body weight, the force on the hip joint is four times body weight.* The joint force indicated in Figure 2-47 is large, 200 to 400 kiloponds (400 to 800 pounds) or more, and so the femoral neck must withstand a large bending moment when the resultant is not along its axis.

The "tension band" approach would dictate pin placement(s) at point(s) (Fig. 2-48A) that would probably be adequate if in addition to the bending moment there were no torques that could cause rotation about the axis of the neck. However, normal walking does cause such torques, and, although they are small compared with the bending moment, they are large enough to make fixation with pins (Fig. 2-48A) inadequate for rotational stability (Fig. 2-48B). Two pins are necessary to adhere to structural stability. (Fig. 2-49).

Even so, the static bending forces at the lateral cortex are too high for the bone. This problem may be met by using a nail-plate combination (Fig.

---

* We have simplified this problem considerably: the weight of the left leg was not subtracted from total body weight; the range of human anatomy contains lever arm ratios larger and smaller than 3:1, and the forces were all assumed to be vertical, whereas the joint and muscle forces are not. A further complication is the presence of dynamic forces, which may considerably exceed the static forces considered here.

13. NAILS, RODS, AND PINS

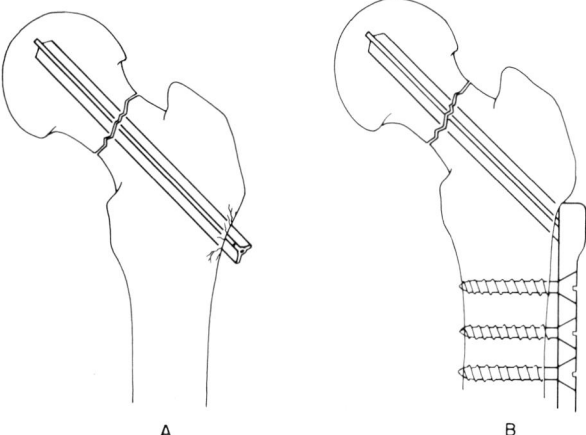

Fig. 2-50. (A) The nail alone results in high forces and possible resorption at the lateral cortex. **(B)** the problem is met by using a plate to transfer and distribute these forces. Since the lever arm(s) of the screws are great, the forces are lower.

2-50). Now, if the bone at the base of the nail resorbs, the bending moment is balanced by forces on the screws in the proximal femoral shaft. Since the screws have reasonably large lever arms, the forces involved are much smaller. These principles are extended further by the arrangement shown in Figure 2-51, the Deyerle pin and plate device.

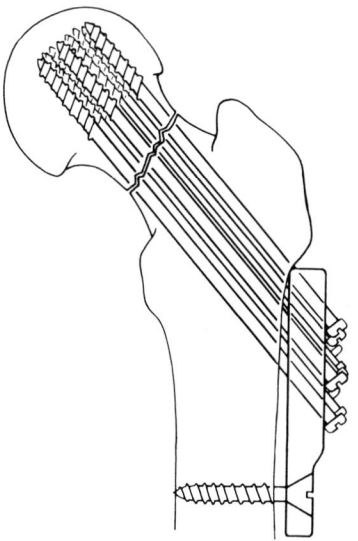

Fig. 2-51. Deyerle pin and plate.

100 PRACTICAL BIOMECHANICS FOR THE ORTHOPEDIC SURGEON

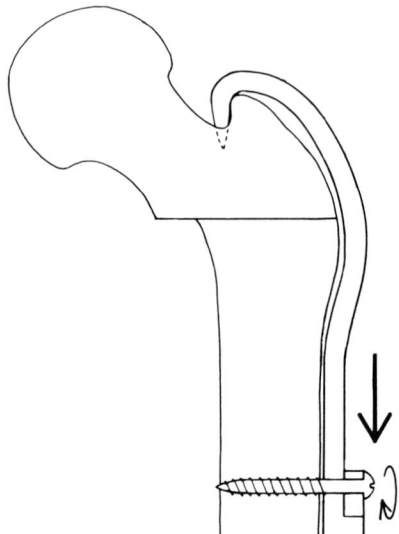

Fig. 2-52. Compression hook to fix osteotomies; the hook is in tension, the osteotomy site in compression.

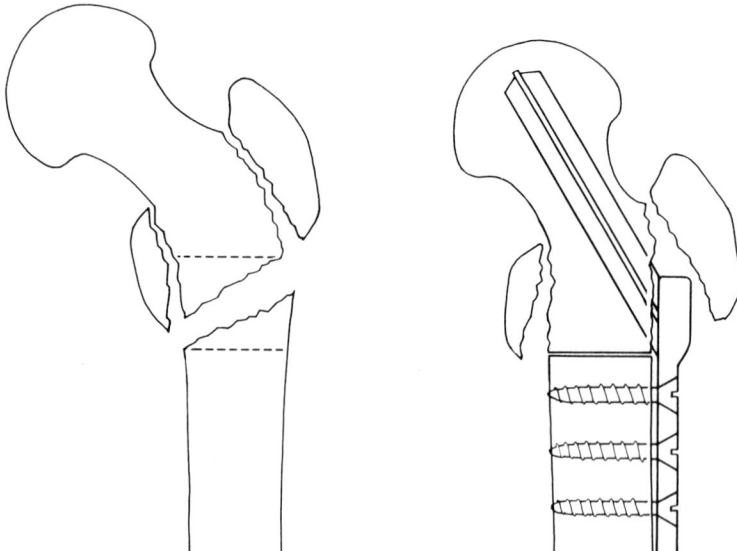

Fig. 2-53. Stabilizing a four-piece intertrochanteric fracture. Note that the femoral head is now in valgus.

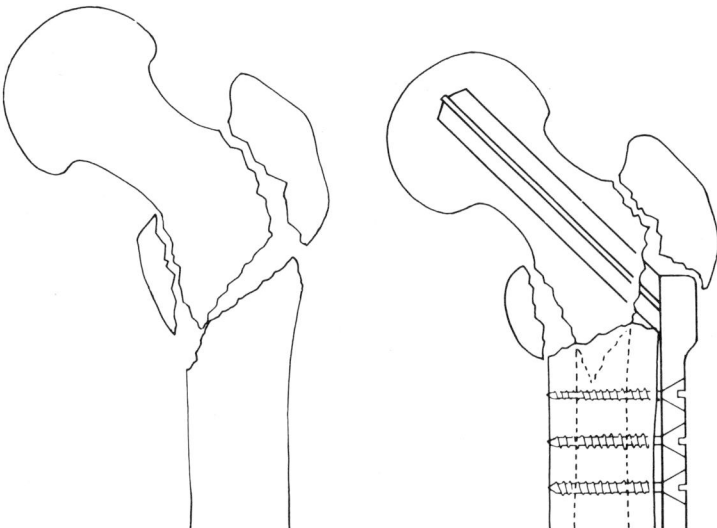

Fig. 2-54. Stabilizing a four-piece intertrochanteric fracture by lateral displacement of the proximal fragment.

The osteotomy compression hook in Figure 2-52 is more of a "tension band" arrangement than the nail-plate arrangements discussed above. The plate has an external hook to engage the proximal fragment and a cortical screw with an eccentric attachment. The fragment is thus compressed against the top of the femur, and the screw assembly is in tension.

None of these devices should be expected to bear normal ambulatory forces alone. Some fail at once at joint forces of 90 kiloponds (180 pounds), when 200 to 400 kiloponds (400 to 800 pounds) are expected (see Fig. 2-47). Thus bone continuity is essential to successful fixation, as the device must fail if nonunion occurs. For instance, unstable four-part intertrochanteric fractures, in which stable, anatomic reduction cannot be obtained, could be best

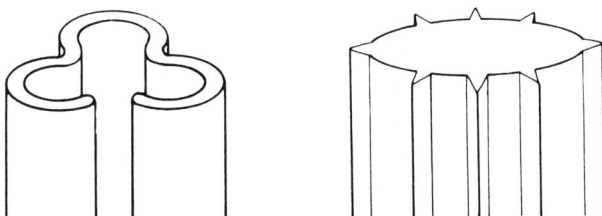

Fig. 2-55. The Kuntschner (clover leaf) rod is flexible but relatively weak in torsion compared with a fluted rod.

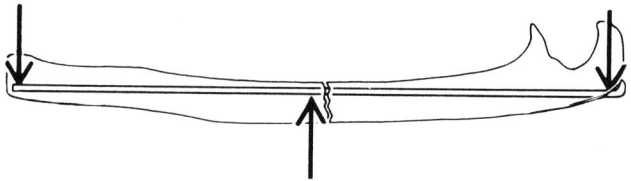

Fig. 2-56. Intramedullary rod fixation by three-point contact with a precurved rod (or straight rod in a curved long bone). Flexibility is necessary for such a fixation.

handled by stabilizing measures such as removing bone or placing the head in valgus (Fig. 2-53) or by displacement (Fig. 2-54).

For simple long bone fractures, intramedullary rods may be used for fixation instead of plates. Inherently, these fixations are not as rigid as plates, since the rod's material is closer to the neutral axis in bending (and torsion).

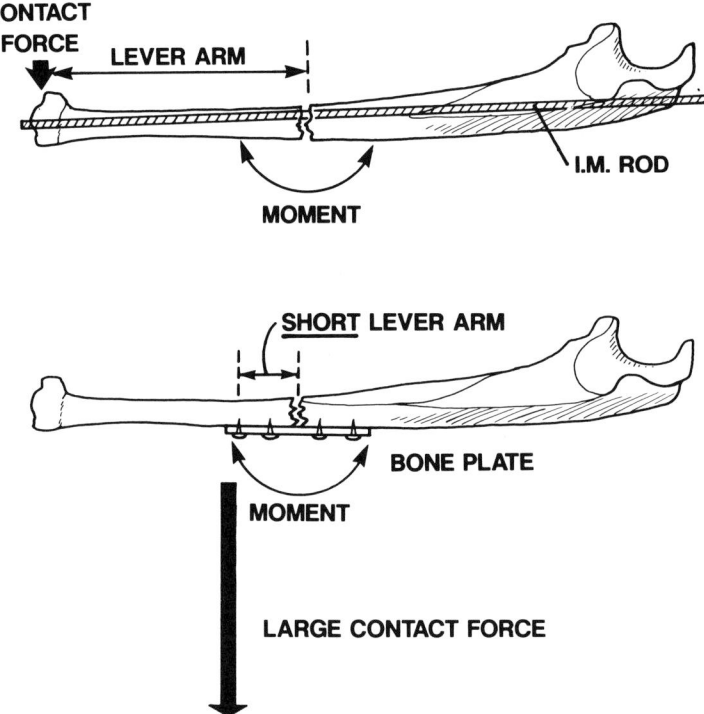

Fig. 2-57. Because of the greater lever arm, much smaller contact forces are required for intramedullary fixation compared with cortical (plate) fixation. An added advantage is the greater flexibility of the intramedullary fixation, reducing or eliminating osteopenia near the fracture.

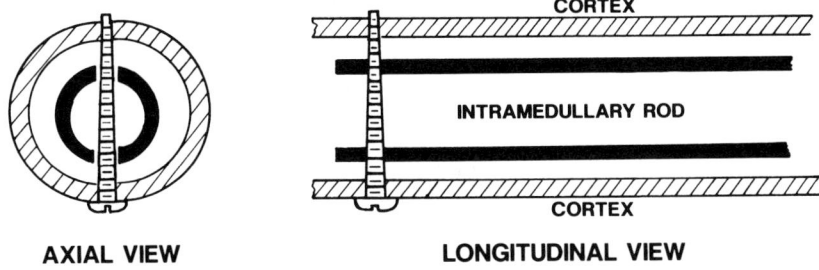

Fig. 2-58. Transverse holes or slots in intramedullary rods (eg, Kuntschner rod, see Fig. 2-55) allow locking by transcortical screws of distal fragments against rotation and sliding.

This may not be an advantage in certain circumstances. If the bone heals slowly, the rod is more likely to break. On the other hand, the increased flexibility ultimately results in less osteoporosis of the bone near the implant, since the bone must bear more of the load. Since intramedullary rods come in a wide range of designs and rigidities, this rationale holds to a lesser or greater degree, depending on the rod design.

The Kuntschner rod, with its cloverleaf cross section, is relatively strong and relatively resistant to bending because of its considerable area moment of inertia. In other words, it has more material farther away from the neutral axis. However, such a rod design is relatively weak against torsion. A fluted intramedullary rod would be much stronger in this modality (Fig. 2-55).

Intramedullary fixation with a curved rod allows three-point fixation. The rod's concave ends press against the endosteal surfaces of the bone above and below the fracture, and the rod's convex apex presses against the fractured area (Fig. 2-56). This arrangement has the advantage of reducing contact forces (Fig. 2-57). The one disadvantage, the tendency of distal fragments to rotate and/or slide, can be eliminated by the use of transcortical screws through slots in the rod ends (Fig. 2-58).

Since long bones naturally curve, it can be argued that straight rods, forced into the medullary cavities, artificially straighten them by angulating the fracture site. The Sage pin was introduced to permit intramedullary fixation of the radius, the natural curvature of which is essential for full forearm pronation and supination. The Sage pin is relatively flexible and thus weak in torsion, and its use must be augmented by external immobilization. The Rush rod—a small, straight intramedullary rod—has few mechanical advantages. Its use should be confined to bone with small medullary canals, such as the fibula. Rush rods are generally too rigid to achieve adequate three-point fixation.

Various practical circumstances can dictate the choice of internal fixation devices; comminuted fractures of the long bone, for instance, frequently

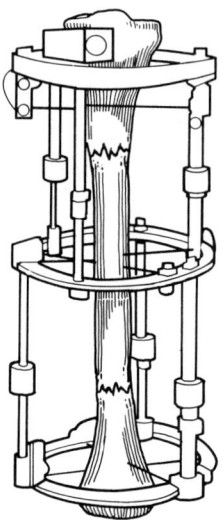

Fig. 2-59. Ilizarov technique. An external circumferential frame is used to hold segments of bone through wires that are strung between attachment points on the frame through the bone.

cannot be reassembled by rods and plates, and screws, or external fixators that hold the fragments in position with wires or pins supported by an external frame (Fig. 2-59), must be used.

14 BONE GRAFTING

A considerable body of experience indicates that bone grafts have the best results when placed on the tensile side and/or near the neutral axis for bending. The idea is to keep the stress on the bone graft as low as possible so that it survives and is incorporated. Consider the "prototype long bone," loaded eccentrically as many long bones are (Fig. 2-60).

Analysis shows that any cross section has a bending moment and a compressive force on it. Total stress distribution is obtained by the simple addition of stresses, as done graphically in Figure 2-61. The magnitude of the tensile stress at the extreme fiber is less than the magnitude of the compressive stress on the other side, since the stress caused by the compressive force cancels some of the stress on the tensile side but adds to the stress on the compressive side. Note also from the figure that the point of zero stress is now no longer at the old neutral axis, but nearer to the tensile side. Thus there is less stress and less movement on the tensile side, which should expedite the establishment of a graft. In this respect, the best location of all is near the neutral axis, but slightly over to the tensile side. The total

14. BONE GRAFTING **105**

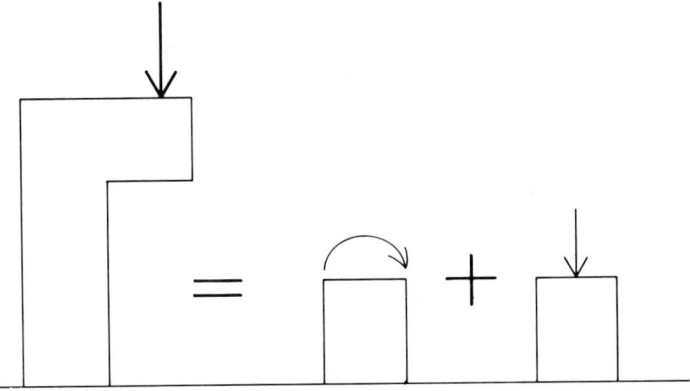

Fig. 2-60. "Prototype long bone" loaded eccentrically (off axis) at the joints. The result is a combination of bending and compressive loading.

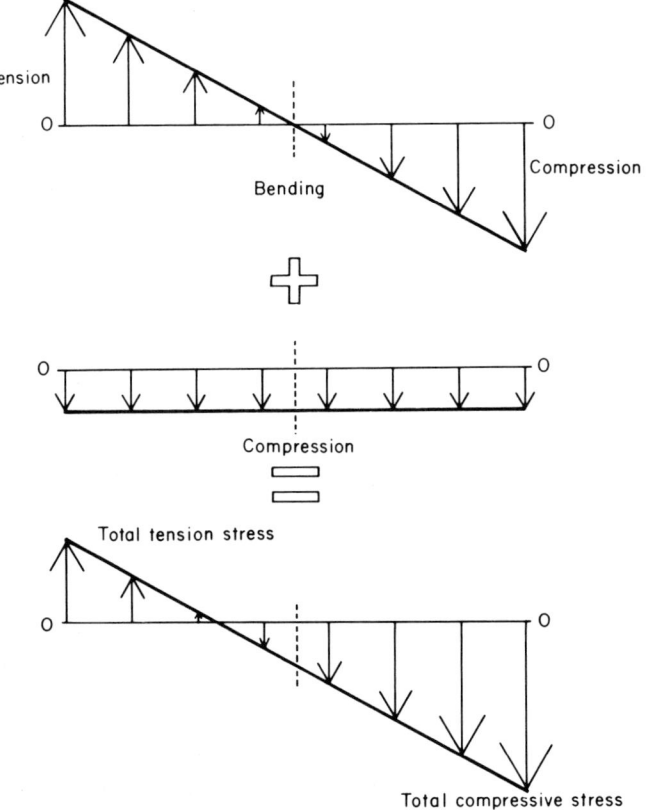

Fig. 2-61. Addition of bending and compressive stress patterns across the bone section results in the total stress pattern shown at the bottom. The maximum tensile stress is lowered.

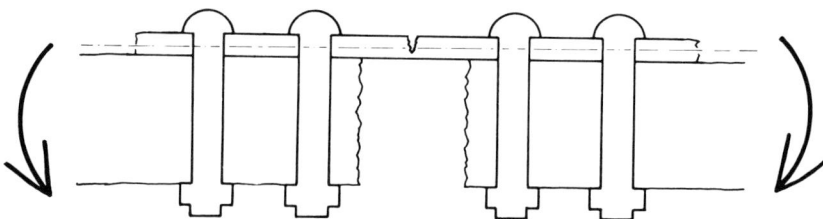

Fig. 2-62. Failure of a graft under bending stresses. If a gap exists, the graft must be reinforced by plates or a rod.

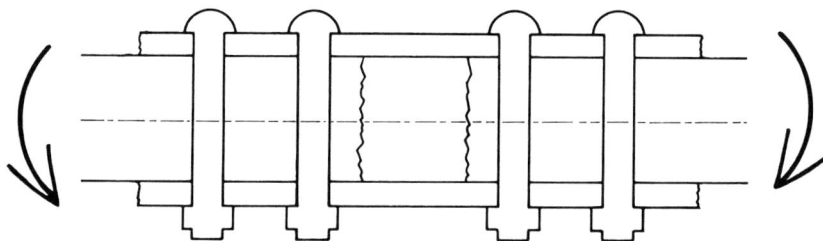

Fig. 2-63. Double grafts bolted across a gap.

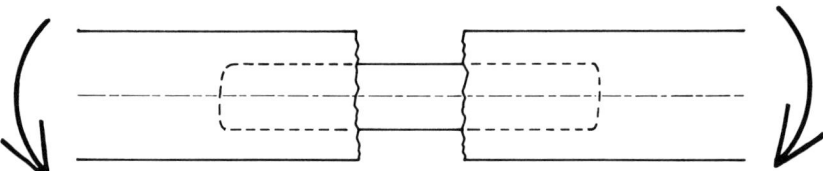

Fig. 2-64. Gap bridged with an intramedullary graft.

stress (bending plus compressive load) is zero or nearly so at this location. The bone graft itself does not perform a structural function, but serves rather as a substrata for the growth of new bone.

In cases where a gap is at the fracture site, it is necessary first to first stabilize the fracture with either double plates or an intramedullary rod before bone grafting. Otherwise the graft must perform a structural function and will probably fail (Fig. 2-62). An alternative is to bolt double grafts across the gap (Fig. 2-63). This is generally difficult because the fracture ends are usually osteoporotic and do not hold bolts well. More practical is the use of an intramedullary fibular graft in the neutral axis, which is thus subject to little stress, while at the same time it acts to stabilize the fracture site (Fig. 2-64). Eventually, the new bone must grow in response to stress;

but the initial graft does best if situated at the point where stresses (and, therefore, strains and relative motion) are lowest.

15 RECAPITULATION

Bending and torsion cause tensile stresses on the long bones. Most forces on the skeleton are from muscle action. These muscular forces on the skeleton can be huge. Stresses on bone can be reduced by articulations, bone shape, and appropriate muscle action. Torsional force can be reduced by articulations. Tensile stresses are required for fracture initiation and propagation. Stress concentrations are an ever present menace. The phenomenon of fracture requires both adequate mechanical energy and sufficient stress. Fracture occurs by either single overload or fatigue. The principles of fracture fixation are tension banding and intramedullary rodding. Intramedullary rods are highly effective because they remain relatively rigid even with a gap at the fracture site. They contribute proportionally less to flexural and torsional rigidity as the bone heals and are associated with relatively small contact forces owing to their relatively large lever arms.

3

Biomechanics of Sports Injuries

1 SPORTS INJURIES AND NEWTON'S THIRD LAW

Over the past few years, since statistics on sports-related injuries have been kept, the number of serious musculoskeletal injuries has been increasing. No practicing orthopedic surgeon needs to be told this: the evidence of this "epidemic" confronts him almost daily in his office. The most dramatic examples are severe cervical spine injuries, particularly from football, rugby, lacrosse, and hockey.

Musculoskeletal elements are torn or broken as a result of high force. In the preceding chapters we discussed force acting on a structure in terms of the stress generated. From a mechanical point of view sports involve the coordinated and skillful manipulation of the body and body segments to achieve a well-defined aim, such as running faster, jumping higher, or propelling a ball. In all of these activities we either move ourselves or move an object. Now we must consider the rate of travel or speed with which a body segment is moving.

The *momentum* of a moving object, its mass times its velocity, provides tremendous force compared with that generated by the same object when stationary and accounts for the greater forces associated with collisions of more massive bodies. Velocity is the rate at which an object is moving. Initiating or increasing this rate of motion is *acceleration*, which requires energy. To do this we become short of breath and sweat, and our muscles become tired. We are imparting metabolically generated energy to create acceleration, thus transforming *metabolic energy* to energy achieved by the motion of an object: *kinetic energy*. Kinetic energy is quantitatively one-half the mass of an object times the square of the velocity the mass is moving. Thus kinetic energy is associated with a moving extremity (or anything else!). The faster we accelerate the limb the more *power* (work or energy per unit of time) it possesses. This power allows us to do something with that limb. In all circumstances, the faster the athlete or his body segments are moving, the greater the energy produced.

In sports injuries, masses that have achieved considerable momentum as they are powerfully thrust through space collide with other masses, in the form of another player, the ground, the bottom of a swimming pool, a moving ball, or a glove, bat, or racquet held or attached to a body segment. In any case a rapid, almost instantaneous *deceleration* of some object occurs. Kinetic energy is transferred, dissipated, or converted to other forms of energy. This decrease in the rate of motion of a mass requires just as much energy as did the initial acceleration.

Remembering Newton's third law from Chapter 1, Section 3 (for every action there is an equal and opposite reaction), we can see that what causes injuries in sports is the equal and opposite reaction either within the limb segment from a collision or from the reaction force created by the acceleration of the body segment itself. For the foot pushing against the ground there is an equal and opposite ground reaction applied to the leg (Fig. 3-1); for the thrown ball there is an equal and opposite reaction applied to the upper extremity.

Reaction force creates kinetic energy within the body segments. The success of the attenuation or dissipation of that energy determines the force and stress levels applied to the anatomic structures. Insufficient shock absorption results in stress concentration and injury.

The effectiveness of shock absorption, the mechanism by which energy is dissipated, is critical in determining whether significant musculoskeletal

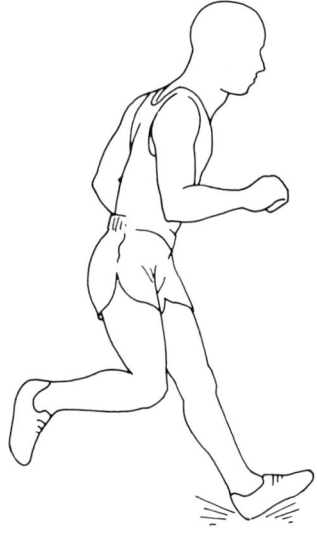

Fig. 3-1. As we run the ground imparts an equal and opposite reaction to the foot.

Fig. 3-2. Considerable force is generated when the head and hands hit the ungiving surface of the bottom of a pool after one dives into shallow water.

injury occurs. Consider diving into shallow water and hitting the bottom of the pool (Fig. 3-2). If the top of your head strikes the unyielding cement floor of the pool, deceleration occurs over a small distance in a short period of time, and fracture dislocation of the cervical spine almost certainly occurs. However, if kinetic energy built up in the process of diving can be dissipated by the body having sufficient time to decelerate in the water before your head strikes, injury will probably not be severe. If the water is deep enough to allow a more gradual deceleration so that by the time you get to the bottom little kinetic energy is left, the dive becomes something you can survive. The important concept illustrated here is that gradual deceleration protects because it dissipates much of the kinetic energy; the momentum of the object is slowed gradually.

Having the necessary distance to slow the deceleration process down enough is a way to avoid injury. We *work* to stop the motion. Work is equal to force times distance. Thus the force related to the work done, or energy required, is inversely related to the distance over which the work is accomplished. The longer the distance over which an object is decelerated, the less the force required to stop it, and thus the less reaction force generated. The extremely short distance for deceleration in the above example therefore implies a large force, hence the spinal damage.

Another mechanism for the gradual dissipation of energy is deformation of surrounding structures. If the pool floor were of thick sponge rubber that would cushion the blow, the deceleration distance would be increased and energy would be used to deform the sponge rubber rather than the skull and neck. Consider the tackling of a charging football player. The kinetic energy of the on-coming player can be substantial (10 to 15 times the energy involved in normal walking), and all of it must be dissipated by forces acting over relatively short distances as the players contact. In head-on tackling (Fig. 3-3) the helmet-protected heads impact with the opposing player.

At first contact the tackled player is traveling at some speed and his kinetic energy is the product of half his mass times the square of the velocity with which he is traveling. To stop him, another player must decelerate him from his initial velocity to a standstill. This deceleration occurs as the contact force between the helmet and the opposing player builds up and presses against the suspension of the helmet and in turn the head of the player acting in the direction opposite the direction of motion of his mass. For the tackler to be decelerated to zero velocity without his gaining much more "yardage," the force times the distance of the decelerating force has to be equal to the initial kinetic energy of the tackler.

In an extreme case, if the tackler runs into a rigid stone wall he is stopped but little cushioning is available from the wall and extremely high forces are generated. In the usual case, compliance is introduced in several loca-

Fig. 3-3. In sports-related activities such as tackling running players, the rapid deceleration of the runner requires substantial force.

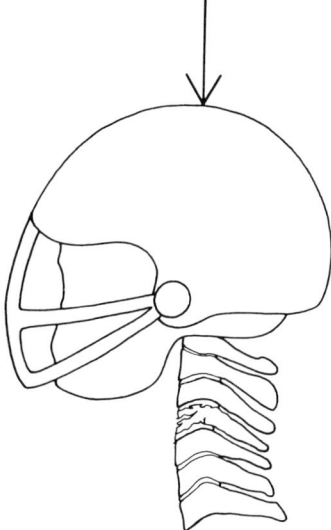

Fig. 3-4. Impact loads directly in line with the cervical spine are likely to cause compression fractures.

tions, including a falling away of the player being tackled, which allows energy to be absorbed over a longer distance, thus reducing the resulting forces. Other cushioning effects come not only from the protective compliance built into the football helmet but also from the compliance of the player's head, neck, other tissues, and bone—and as soon as his shoulder touches the compliance of the shoulder pad of the tackling player. Whether the cervical spine is injured depends crucially on the type of motion allowed when the decelerating force is applied.

Figure 3-4 generally indicates an impact force on the top of the head centrally in line with the cervical spine. Since the impact force is directed along the axis of the spine, no bending moment is introduced that would tend to flex or extend the spinal axis from its indicated direction. The compliance in this situation arises strictly from the compression of the helmet and its suspension, the skull, vertebral bodies, and discs. Muscles can play only a stabilizing role in this situation since they cannot resist compressive forces.

The loading situation changes significantly when the tackle contact or other impacts to the head are from a nonaxial direction such as a frontal face mask impact or impact from the back of the head, causing either extension or flexion of the cervical spine, respectively. In these situations the spine is subjected primarily to a bending mode of loading with general tension generated at the convex surface and compression at the concave surface of the deforming spine. The bending moment acting in any section of the

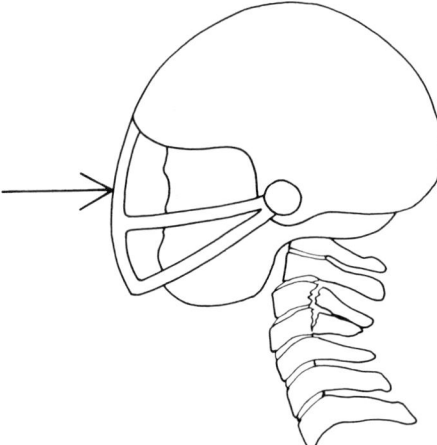

Fig. 3-5. Impact loads at some angle to the cervical spine are likely to cause flexion or extension injuries.

spine is proportional to the moment arm through which the impact forces act. Since the inherent resistance to flexion-extension is rather low, the impact force can travel through a considerable distance before, say, the spinous processes limit further extension and a typical hyperextension fracture of the intervertebral spaces is produced (Fig. 3-5). In this eccentric loading situation, under certain circumstances, paraspinal cervical soft tissue muscles can act to reduce the "shock," as discussed in Section 2.

When impulsive loads to the head tend to bend the cervical spine, the stabilizing ligaments and muscles of the spine can play an extremely important role in absorbing the energy of the impact if considerable deformations can be tolerated. The ligaments and muscle tissue on the tension side of the spine require a stretching force to produce elongation, and these forces times their displacement create work at the expense of the impact energy. The tensile force generated in the muscle and connected tendons are transformed into compressive forces in terms of the vertebral bodies and discs. Thus, although the tensile strength of the vertebral bodies is quite low compared with their compressive strength, the geometry of the tendon and muscle containment of the spine together with the available large motions act to apply compressive rather than tensile force on these skeletal members. Each component of the musculoskeletal system that deforms under the applied load is effectively reducing the energy that has to be dissipated by the remaining components and thus reduces the peak felt by any component of the system.

As the shock or peak dynamic load is created by sudden accelerations or decelerations, the problem of reducing peak accelerations to acceptable levels boils down to increasing the distance over which the energy is dissipated, or arranging a more compliant energy reception. When we accelerate an arm to throw or hit a ball, after the ball has been thrown or hit we "follow through" with further arm motion. This dissipates energy and slowly stops the arm. Failure to "follow through" properly can cause injury. Deceleration involves a lowering of velocity with concomitant dissipation of such energy and depends on the distance over which we can decelerate.

Another way actively to diminish the impact loads on the skeletal system is to avoid them. If we can diminish the amount of kinetic energy by minimizing the acceleration of our body segments at the instant they interact with unyielding surfaces, then we can diminish the amount of kinetic energy that needs to be dissipated.

Vertebrates do this by unconscious control of their movements. One can detect, in most subjects, noticeable efforts to decelerate the body parts before they strike an object. For example, at heel strike, most of us slow the acceleration of the lower limb and decelerate the heel so that its kinetic energy is less at the time of impact. These reactions are faster than spinal reflexes and are controlled by the so-called motion generators in the brain stem. Not much is known about motion generators except that they are pre-programmed and are important coordinators of all activities. For example, almost everyone uses a hammer in the same way and has certain common characteristics of gait.

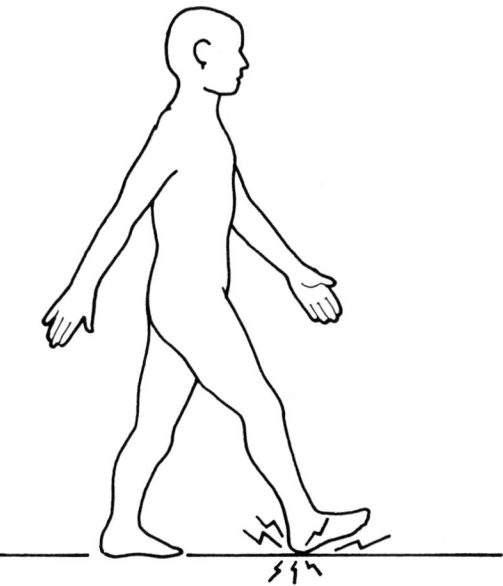

Fig. 3-6. Failure to decelerate the limb just before heel strike will subject the musculoskeletal system to impulsive loading.

116 PRACTICAL BIOMECHANICS FOR THE ORTHOPEDIC SURGEON

Fig. 3-7. Catching a baseball with a catcher's mitt.

A small but significant percentage of humans appears to have aberrations in the ability to decelerate body segments. Since this deceleration must occur over a few thousandths of a second, timing is crucial. An untimely deceleration, particularly one that comes too late, will not help. We refer to individuals who seem to have difficulties with untimely decelerations as *microklutzes*. Microklutzes repeatedly subject their skeletons and joints to repetitive impulsive loading. Chronic microklutziness leads to fatigue damage of articular and bony structures (Fig. 3-6).

A simpler example that illustrates all of the above is the catching of a powerfully pitched baseball with a catcher's mitt (Fig. 3-7). As the ball strikes the mitt some of the ball's kinetic energy is dissipated by the creation of soundwaves (the pop of the ball hitting the mitt), heat, and deformation of the material of the mitt itself. But if the mitt is held steady, most of the ball's kinetic energy has to be dissipated over the distance the mitt deforms. Trying to catch a pitched baseball in this fashion results in a bruised palm and possibly even a metacarpal fracture (Fig. 3-8).

Successful catchers decelerate the ball by moving the hand in the same direction the ball is traveling (Fig. 3-9). Thus you can generate the energy required to decelerate and stop the ball with a small force since the force required is multiplied by the distance to achieve the energy transfer. This explains why if you move the mitt back as you catch the ball your hand does not sting. One allows the hand to retract with the ball and decelerate it slowly, resulting in small contact forces.

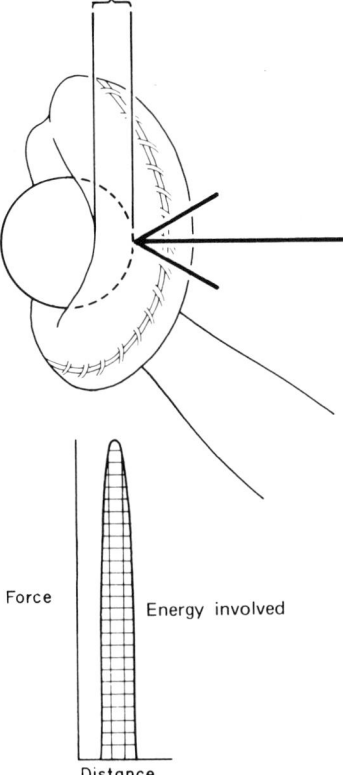

Fig. 3-8. Catching a powerfully thrown baseball in a catcher's mitt without moving the hand decelerates the ball in the distance the glove deforms. This dissipates the kinetic energy of the ball over a relatively short distance, thus generating substantial peak force.

The actual peak force developed during such energy absorption is a direct function of the compliance of the system under the applied load, in other words, how much it moves. Note that in Figures 3-8 and 3-9 we have plotted the peak forces developed. The areas under the curves represent the energy involved, and the areas under both curves are equal because it takes the same amount of energy to stop the ball whether we bring our glove back or not. That amount of energy is determined by how fast the ball was traveling and by its mass. But if we decelerate over a distance we can dissipate the energy with a small force and thus spare the tissues of our hand.

The bony shafts, much as hardwood, bend and deform under applied loads. This bending can act to absorb energy, as can the deformation of the soft tissues of the palm. But in catching a powerfully thrown ball with as little contact force as possible, most of the compliance must be produced by muscle action and joint motion, which bring the hand and glove back, gradually decelerating the ball.

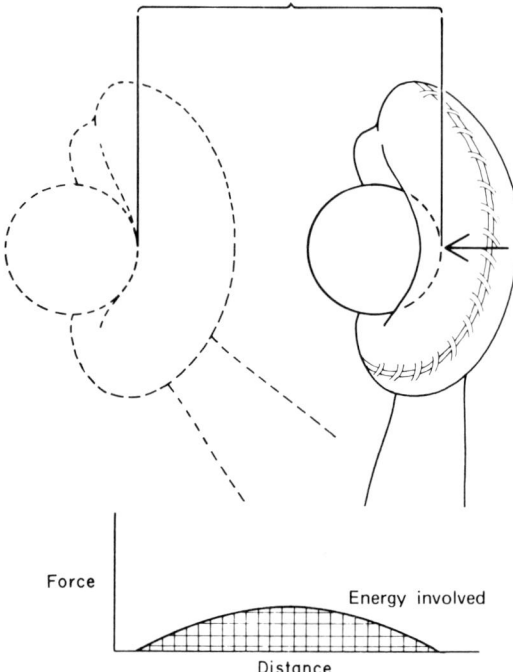

Fig. 3-9. As in Figure 3-8, except that the hand is moved back as the ball is caught. The energy is dissipated over a relatively long distance, and the peak force developed is substantially reduced.

Enhancing musculoskeletal shock absorption protects athletes from injuries. Helmets that are suspended on the head and allow space in which to deform before crushing into the skull, compliant pads over exposed bony prominences and relatively compliant playing surfaces, and shoes that afford traction but do not rigidly fix the foot on the ground so that the lower extremity can buckle on impact all help.*

Snowmobiles fitted with shock-absorbing seats would probably lower the incidence of the compression fractures of the spine that result from the skimobile dropping into ruts and ravines at high speed. There is no way in contact sports to avoid collision, and it is clearly the unprotected and unprepared for collisions that cause injury. Rule changes designed to prevent "blind side" tackles and better designs of protective equipment are goals orthopedic surgeons should promote.

---

*The biomechanical advantage of short skis with quick release bindings is obvious. While still attached to the foot, skis act as lever arms multiplying rotational bending forces on the lower extremity. Short skis decrease this lever arm.

2 ACTIVE MUSCULOSKELETAL SHOCK ABSORPTION

We have discussed in the preceding section that, besides protective equipment, active control of limb segment movements has a significant effect on lowering the impact forces felt as we try and stop a rapidly moving object. Properly timed and executed muscle contracture, besides moving parts of the body, can also act in and of itself as a substantial shock-absorbing mechanism.

Most of the time we can jump from a reasonable height without severely injuring ourselves. The high jump and broad jump are not particularly dangerous activities. Besides the compliant sand pits or rubber mats we land on, the mechanisms that spare this shock loading are reflexural in origin and are neuromuscular. Most of us can land on a hard surface from a 2 m (6 foot) height without injury. The impact of the body hitting the floor is primarily absorbed when we straighten the hips and knees and dorsiflex the feet (Fig. 3-10). To do this we have to land with the hips and knees bent and the ankles plantar-flexed. Getting them into position on the way down, before we land, is a reflex action.

This extension of the limbs after impact, if totally controlled by muscles, stretches them. The stretching of muscles requires work. Think of the muscles as rubber bands. What the body seems to do is to place the joints into

Fig. 3-10. We protect ourselves from impulsive loads when we land from a jump with our lower extremities flexed

position and by reflex contracts the muscles so that as we land the joints move and their associated muscles are allowed to stretch against some resistance. As we land, the energy of the body hitting the floor is primarily absorbed by flexing the knee and pulling on the contracted hamstrings and gastrocnemius muscles. The "stiffer" the landing, the smaller the deformations and elongations to these muscles, the higher the impact forces.

When we cannot move and stretch the muscles adequately because we are unprepared for the load, as in unexpectedly stepping off a step or when the impulsive load is applied to the long common axis of two limb segments (as if the knee were straight during heel strike), the shock loading has to be completely attenuated or transmitted by the compliance of the loaded skeletal elements themselves: the bone shaft, the cartilage, and whatever soft tissue is in line with these elements. In such a situation muscle cannot contribute substantially to energy reduction because inadequate joint motion is generated to invoke significant reflex action and muscle contraction. In jumping from even small heights, a stiff-knee landing can be truly painful.

Thus, controlled joint motion on impact provides several shock-absorbing (energy-dissipating) mechanisms. The movement (rotation) of the joint also acts as a dissipater, providing the system with redundant or wasted motion. Muscles act to allow the limbs to decelerate over time, lowering the peak dynamic forces generated on deceleration. Finally, the active stretching (by segment motion) of muscles under slight tension absorbs vast amounts of energy. One of the secrets of avoiding injuries from impact is to be prepared.

3 SPORTS INJURY RESULTING FROM FATIGUE

If the mechanism of skeletal shock absorption by deformation of tissue (and muscular contraction) breaks down because of an abnormal loading situation, or if the total energy of impact exceeds the capabilities of the shock-absorbing mechanisms, fracture of bone and/or rupture of soft tissue results. But damage can result from repeated shock loading, even though each load is well under the threshold the joints and other musculoskeletal tissues can tolerate.

Consider running or hurdling: normally the knee, because of its large range of available flexion-extension and surrounding strong muscles, acts as an extremely good shock absorber because the knee flexes on impact, distributing the required deceleration over a longer distance and absorbing most of the energy by stretching the hamstrings and gastrocnemius muscle (Fig. 3-11). If the runner or hurdler overextends himself, his muscles fatigue and no longer function at just the right moment to allow them to be stretched under tension. Muscles not contracting at the right moment get overstretched and are strained or sprained. More importantly, the kinetic energy

Fig. 3-11. Most of the impact from jumping a hurtle is absorbed by stretching the posterior leg muscles.

of impact remains to be absorbed to a large extent by deformation of the skeletal tissues themselves. Repeated impact of this nature, although well below the fracture threshold, can produce cumulative damage.

The failure of bone in fatigue is mentioned in Chapter 1, Section 12, and Chapter 2, Section 5. We have also introduced the concept that external loads on the skeletal members are always eccentrically applied, thus primarily creating bending and compressive stresses within the bone and tending to tilt one bone on the other at the joints. Bone is weaker in tension than in compression, and tension is the mode of fracture initiation in bone. Clearly, then, lowering the tensile stress minimizes the incidence of fatigue fracture of bone to which the athlete would be particularly prone because of the repetitive nature of sporting activities.

Experience demonstrates that it is not the well-trained athlete who is subject to the fatigue (march or stress) fracture, but rather the "Sunday" athlete. The reason for this is that muscles play a significant role in lowering of the bending stress in bone. Consider the principle of a guy wire used to hold up a radio aerial or telephone pole (Fig. 3-12). Here the bending stress is lowered and therefore the tensile stresses are significantly reduced (see Fig. 2-57). In a similar manner, the iliotibial band lowers the tensile stress in the femur (Fig. 3-13).

The body weight acting centrally tends to bend the femur, creating tensile stress along its lateral edge. The iliotibial band, with its tension controlled by the tensor fascia lata muscle, acts as a tension band and applies a compressive force laterally along the femur, significantly lowering the tensile

122 PRACTICAL BIOMECHANICS FOR THE ORTHOPEDIC SURGEON

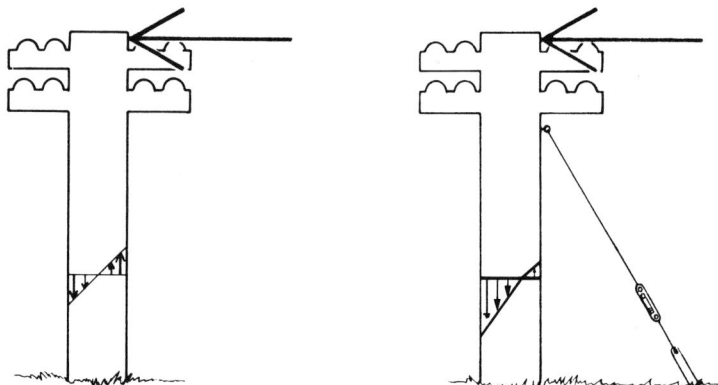

Fig. 3-12. The tensile stress created by the attempted bending of a telephone pole can be minimized with a guy wire. Note that the tensile stress is reduced because the guy wire acts to keep the pole from bending. The guy wire helps to pull the pole further into the ground. Thus the compressive stress on the pole is increased.

stresses on that side of the bone. The addition of compressive stress in exchange for a lower tensile stress is a fair trade-off, since bone is relatively strong in compression.

Consider the combined effect of the short and long head of the biceps and brachialis in reducing the bending stress in the humerus and forearm (Fig. 3-14).

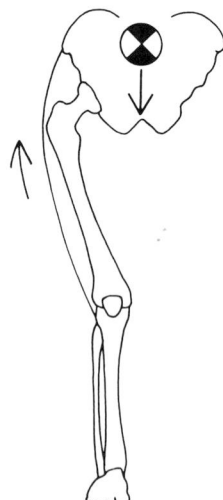

Fig. 3-13. The iliotibial band acts as a guy wire, reducing the bending stress on the femoral shaft but increasing its compressive stress.

3. SPORTS INJURY RESULTING FROM FATIGUE   123

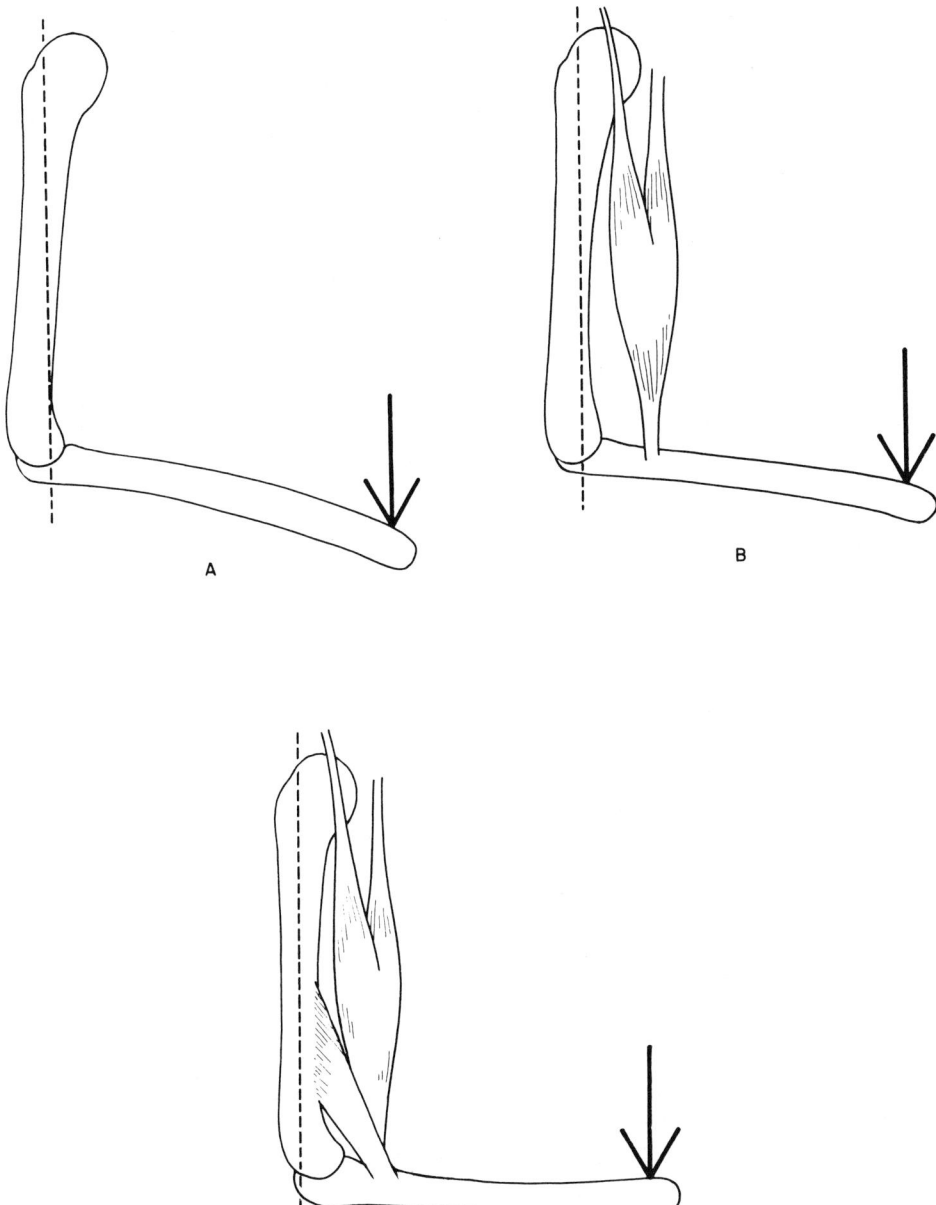

Fig. 3-14. (A) Without a biceps of brachialis, the humerus and forearm bones take the full bending stress caused by the external force. **(B)** The biceps acts as a guy wire, reducing the bending stress on the humerus and proximal forearm. **(C)** The brachialis reinforces the anti-bending action of the biceps.

One of the major roles of muscle, besides ensuring torque equilibrium and producing motion about joints, is to minimize the tensile stress in bone. Thus fatigue of muscle in the poorly trained athlete creates significant tensile stresses in bone that can result in stress or fatigue fractures. These cracks are essentially an accumulation of fatigue damage at loads below the forces that would normally create fracture.

There is some indication that bone undergoes a continuous process of local damage and rehealing. Because of the nonhomogeneous properties of bone, at a very local level stress concentrations can occur in the bone structure under loads that, although inadequate to damage the overall structure, can produce microfractures. These fractures essentially serve to absorb part of the energy that the overall structure managed to absorb without gross damage, and they relieve the stress on this portion of the overall structure so that continuous application of loads to the overall structure will not lower other portions.

The microfracture, if trabecular, forms a callus that later resorbs, thus healing the local defect. If cortical, the microfracture heals by remodeling. This process appears to be stable at subcritical loads. However, repeated loads exceeding this threshold can produce cumulative microfractures that then accumulate and form stress fractures.

There are also indications that this microfracture mechanism for energy absorption impact of loading naturally occurs in the subchondral bone of arthrodial joints, and then an imbalance of the stable equilibrium between microfracture and local damage may lead to joint deterioration. Healing of the excessive numbers of microfractures in the subchondral bone can produce local changes in bone stiffness that then create stress concentrations as the load from the cartilage is distributed throughout the subchondral structure. More is said of this in Chapter 4.

4 MECHANICS OF LOCOMOTION

Most sports involve running. Even for the well-trained athlete short bouts of sprinting can be exhausting, whereas walking for prolonged periods, even in the untrained person, can be accomplished without fatigue. In running we increase our kinetic energy, deriving most of it from an increased metabolism. In walking we minimize the contribution to kinetic energy from metabolic sources. How is this accomplished?

Whenever a mass is elevated, gravity tends to make it fall from its height. Although it takes energy to lift the mass, it falls without the active expenditure of energy, and in fact kinetic energy increases. A mass, which has the potential to fall and thus has gravity act on it, can be considered to have *potential energy* that is equal to the energy required to lift the mass and also equal to the kinetic energy at the end of the fall. In other words, the work done on the mass against the force of gravity is potential energy. If a person falls, this work is converted to kinetic energy.

Fig. 3-15. A rollercoaster requires kinetic energy to get it to the top of its track. The ride down requires no energy; energy has to be expended to stop it.

Consider a roller coaster (Fig. 3-15). For it to move initially from a stopped position to the top of its track requires acceleration, provided by an engine or motor of some sort. Kinetic energy is expended. As the roller coaster gains altitude it develops potential energy, proportional to the height achieved. At the top of its track it has developed maximal potential energy, and the transfer of kinetic to potential energy is complete. The ride down requires no further kinetic energy. The increased acceleration of the roller coaster on the way down is the result of potential energy; as it gains acceleration, it transfers potential energy to kinetic energy. At the bottom this transfer is complete. In an idealized frictionless environment the roller coaster could just keep on going in the same direction at a constant velocity without the need of additional energy to keep it moving. Frictionless situations never exist on earth, however, and certainly not on roller coaster tracks. In reality one has to apply energy to maintain constant velocity except in outer space.

Potential energy is the quantity of the mass raised multiplied by the height to which it is raised (Fig. 3-16). In discussing changes in the situation of the mass it is easier to think about that mass as if it were concentrated in one point. This point is called the *center of gravity*, and it is the spot where one would place a pin in order to balance that mass on the pin. In humans standing at attention, the center of gravity is just in front of the second sacral vertebra. With changes in the positions of the limbs and head, as in gait, the center of gravity shifts.

For simplicity of analysis we can describe the motions of the body through space in gait as changes in the position of the center of gravity. A good description of human gait is that we basically elevate our center of gravity, allow it to fall, decelerate it, and then accelerate the center of gravity back up. We repeat this process over and over so that we continue walking rather than falling down. Just standing endows the human center of gravity with

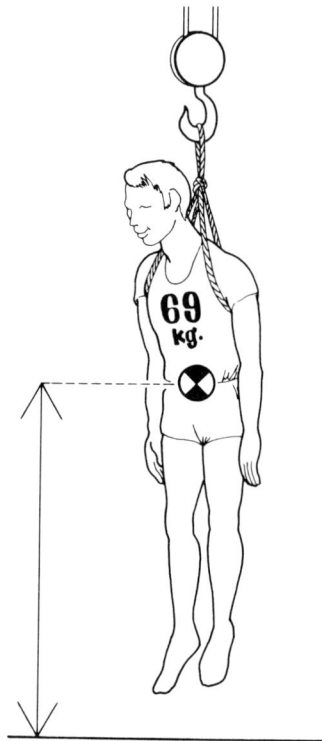

Fig. 3-16. Potential energy depends on the mass of the object and the height to which it has been raised. For convenience, the center of mass (center of gravity) is used for these measurements (see text).

significant potential energy. To lift the center of gravity up, work is done in addition to the work necessary for forward acceleration (Fig. 3-17).

Potential energy depends on the position of the center of gravity. It is at its highest at the mid-stance phase of gait (Fig. 3-18). We use potential energy in gait to fall forward and achieve a relatively metabolically free ride through part of the gait cycle. The "falling down" part is important as it converts potential energy to kinetic energy, with little or no additional work necessary.

We build potential energy by kinetically elevating the center of gravity. The main reason human gait is so efficient in its expenditure of energy is that we achieve a trade-off between kinetic and potential energy. The center of gravity moves up during the beginning of stance or swing phase (Fig. 3-19) and comes down just after the middle of the stance or swing phase. There is thus a trade-off of potential for kinetic energy in gait. At just the time the kinetic energy demands are greatest, our potential energy is maximal

4. MECHANICS OF LOCOMOTION   127

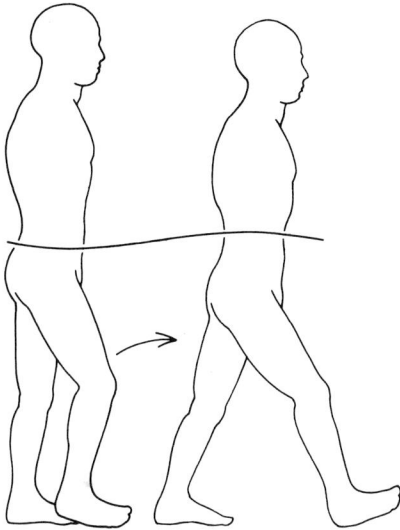

Fig. 3-17. As we swing through in gait, the movement of the swing leg consumes kinetic energy. Actually most of it is spent decelerating the leg. As a result, kinetic energy is at its peak in the swing phase of gait.

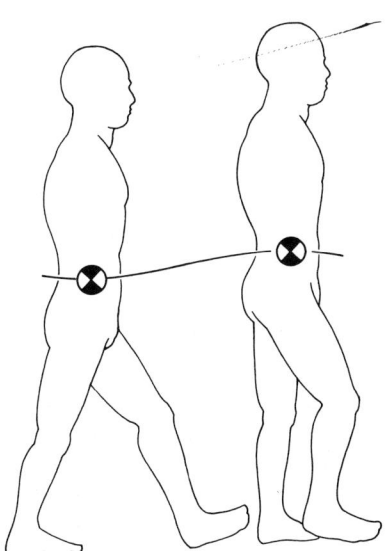

Fig. 3-18. As we move over the stance foot we elevate the center of gravity of the body, thus increasing our potential energy. Potential energy is thus greatest in stance and can be maximally utilized in the swing phase as the center of gravity drops.

128 PRACTICAL BIOMECHANICS FOR THE ORTHOPEDIC SURGEON

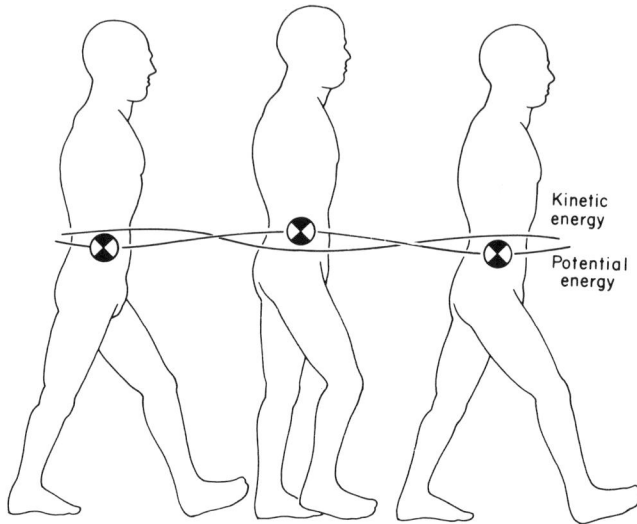

Fig. 3-19. Note that, in gait, potential energy is maximally available when the demands for kinetic energy are greatest. A trade-off can thus occur, lowering the demands for kinetic energy.

and we can trade it off. The additional kinetic energy demands in gait are therefore minimal.

Kinetic energy is produced by metabolic processes. Most of the metabolic energy expenditure in gait is not used to accelerate the center of gravity but rather to decelerate the inertia created by body motion. Once you "get going" it takes relatively little energy to "keep going."

Clearly we want to minimize energy expenditure, and in forward motion we should not waste energy by an unproductive movement of the center of gravity. Some vertical displacement is of course necessary to achieve a potential energy–kinetic energy trade-off. But too much up and down motion in gait would be wasteful. You can demonstrate this by bouncing up and down at every step and seeing how quickly you tire. There are certain coordinated joint motions that we instinctively use to limit vertical displacements of the center of gravity. These are rotation of the pelvis, which allows us to move one leg ahead of the other with a minimum of excursion of the center of gravity; pelvic tilt or dropping of the pelvis on the swing side, which further limits vertical elevation of the center of gravity because it keeps the swing half of the body lower than you might expect; and knee flexion on the stance side, which tends to shorten the stance limb relatively and prevent unnecessary upward motion of the center of gravity (Fig. 3-20).

The coupling of knee extension with ankle dorsiflexion and knee flexion with plantar flexion maintains a relatively equal length leg throughout the stance phase as we come down initially on our heel, then transfer our weight along the foot and roll the foot off the ground by coming up on our toes (Fig. 3-21). The relative adduction of the femoral shaft with physiologic valgus

4. MECHANICS OF LOCOMOTION   **129**

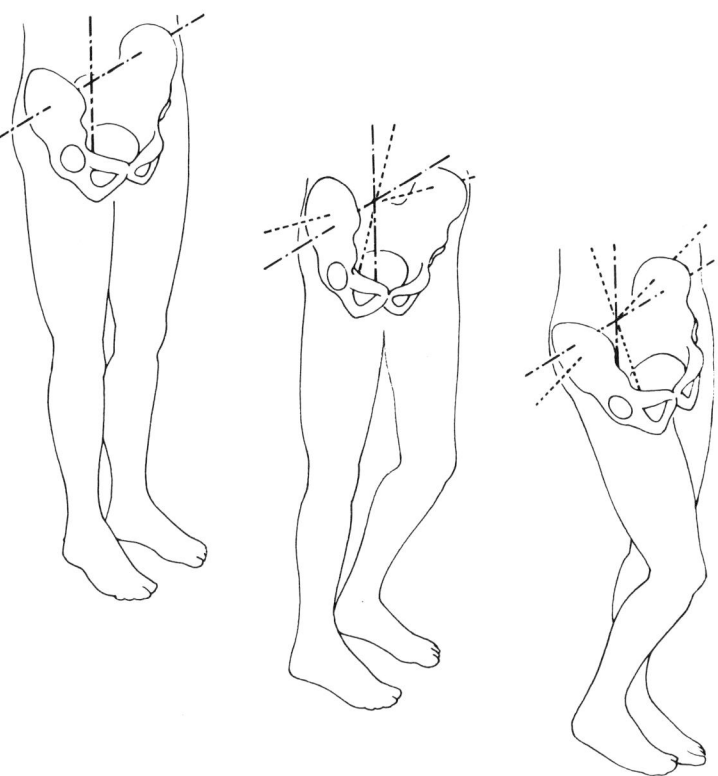

Fig. 3-20. Rotation of the pelvis, pelvic tilt, and knee flexion on the stance side all act to minimize excessive vertical motions of the center of gravity during gait.

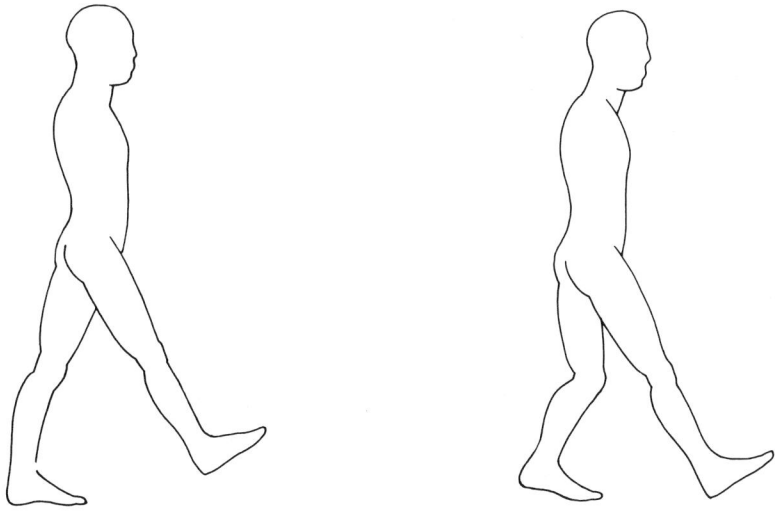

Fig. 3-21. Coupling of knee and ankle motions helps to maintain a level center of gravity during gait.

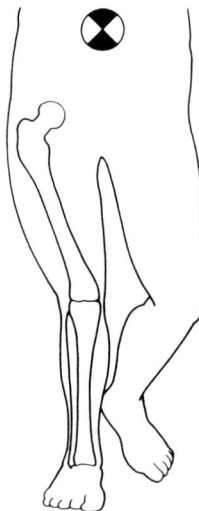

Fig. 3-22. Physiologic valgus of the knee minimizes lateral displacement of the center of gravity during gait.

of the knee permits the foot to be placed close under the center of gravity without its excessive lateral shift in the stance phase (Fig. 3-22). Also, planting of the feet is well coordinated during gait. The double support phase of normal walking lasts about 15% of the gait cycle.

In running, the double support phase is eliminated. There is also a much greater vertical displacement of the center of gravity as the legs are used for forward propulsion and there is a substantial upward component of force created in this motion. The lack of a double support phase also requires a wider based foot placement for balance than does walking. In running, therefore, kinetic energy is increased to such a great extent that the moment of potential energy gained with the larger vertical movements of the center of gravity is insignificant compared with the tremendous amounts of kinetic energy required to keep running.

5 MECHANICS OF MUSCLE RUPTURE, BURSITIS, TENDONITIS, AND MENISCAL TEAR

In previous sections we stressed how joint motion and muscle lengthening act as shock absorbers and lower the tensile stress on bone. Failures of the muscles and joints in these usually well-coordinated activities are not uncommon, and soft tissue injuries are the most frequent injuries acquired by athletes.

Fig. 3-23. Rupture of the heel cord usually results when the gastrocnemius muscle is stretched while it is being forcefully contracted, such as when a runner unexpectantly steps in a hole.

Muscle ruptures if passively pulled beyond its ultimate tensile stress. Given the limitation of joint motion, muscle rupture is usually a rare injury from simple stretch in the absence of direct trauma to the muscles or joint dislocation. More common is complete rupture of a muscle because of forced stretching while it is in contraction. This results from instant incoordination and is what usually happens when the biceps tendon or heel cord ruptures (Fig. 3-23).

What is more common than complete muscle rupture is partial muscle tear. In the inadequately trained athlete, muscles fatigue. Contractions become insufficient to create the required tension. The athlete keeps trying to perform. Muscle substance is subjected to passive stretch, which creates a small, incomplete tear, commonly referred to as a "charley horse."

The tendon sheaths that envelop moving tendons and the bursae that separate muscles from underlying bony prominences both occur in places where the muscle pull undergoes change in direction. They are subject to trauma at these points. A bursa or tendon sheath is usually located at the fulcrum of a muscle excursion (Fig. 3-24), and in such a place can have considerable compressive force applied to it. Such force can damage the lining of the bursa or tendon sheath, creating inflammation that, if continued, becomes chronic. Such chronic inflammation in a bursa is "bursitis"; in a tendon sheath it is "tendonitis."

Tennis elbow is an excellent example of a chronic tendonitis caused by a rupture of the deep aspect of the common extensor tendon of the wrist and

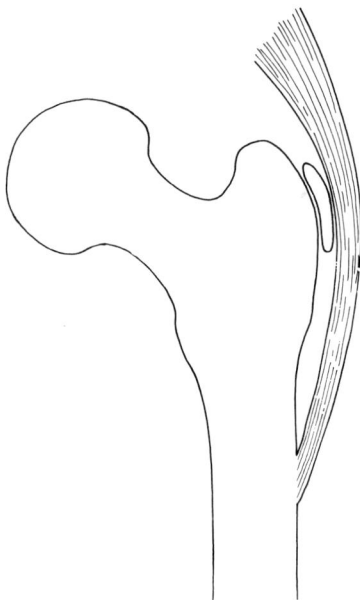

Fig. 3-24. Bursae are in areas where a muscle or tendon rides over a bony prominence.

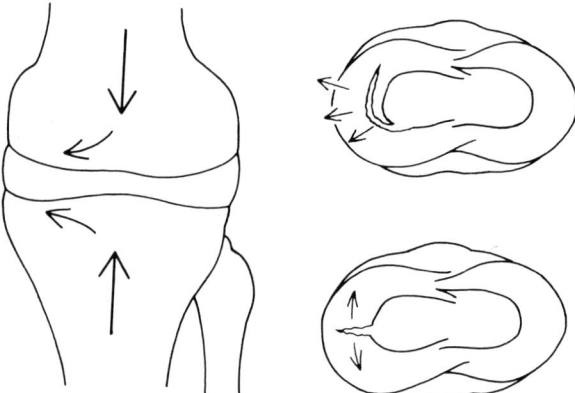

Fig. 3-25. Meniscal tears in the knee result from excessive stress on the semilunar cartilages when full knee motion is blocked.

finger extensors as they originate on the lateral epicondyle of the humerus. Because these extensor tendons are used in almost all hand motions, this area is almost never rested and the sprain becomes associated with inflammation and remains chronic. Attempts to treat tennis elbow mechanically by taking the strain off the extensor tendons with a tight upper forearm band, or a larger grip racquet handle, or a change in stroke, particularly in the back hand, are usually successful.

The most common serious sports-related soft tissue injury is a tear of the semilunar cartilage or meniscus of the knee. Such injuries normally result from a blow to the knee in a situation that fails to allow adequate knee motion to dissipate the energy. This results in extremely high intra-articular forces that exceed the tensile strength of a semilunar cartilage. Knee motion involves a fixed coupling of a flexion-extension with internal and external rotation of the tibia on the femur. If the knee is flexed or extended in such a way that the obligatory concomitant rotation cannot be carried out, the meniscus is caught between the tibial condyle and the femur under a shearing type load that produces an extruding type stress with tension in the direction shown in Figure 3-25.

4

Mechanics of Joint Degeneration

1 OSTEOARTHROSIS AS A WEAR AND TEAR PHENOMENON

Although metabolic and enzymatic factors are involved in joint degeneration (osteoarthrosis), the process is best thought of as the final common pathway of mechanical deterioration of joints, occurring from an imbalance between the stresses applied to the joint and the ability of the tissues to resist that stress. Thus osteoarthrosis can result from either excessive or poorly handled stress or from an inherent structural weakness of the articular cartilage. In the latter situation the mechanical implications are clear. The yield stress of the cartilage is lowered, and even normal stress overwhelms its structural integrity.

In this biomechanical approach we will not discuss joint degeneration associated with underlying metabolic abnormalities; we will discuss only situations clearly associated with a primary mechanical cause. Incongruity from any cause results in stress concentrations within a joint, as does relatively unprotected trauma. If the traumatic forces are severe enough, fracture of a long bone or rupture of ligaments results. Low level, repetitive trauma is most likely to cause osteoarthrosis. For example, pitching a curve in baseball involves a rapid rotatory acceleration or snap of the forearm when the ball is released. The elbow joint lacks effective shock-absorbing mechanisms to protect the radiocapitellar articular cartilage from such a repetitive insult, and osteoarthrosis results. This can occur even in skeletally immature persons and has been the reason behind attempts to ban the curve ball from "Little League" competition.

It is not unusual for the stress on articular cartilage to be 1.4×10^6 to 3.5×10^6 newtons/m^2 (200 to 500 pounds per square inch). Even under such punishing repetitive loading, most joints function well for decades. In this chapter we discuss how joints mechanically function and the role mechanical factors can play in the development and treatment of osteoarthrosis.

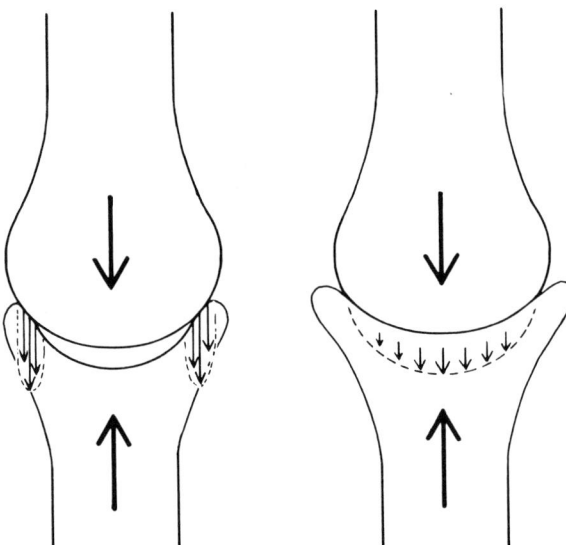

Fig. 4-1. Under load the joint conforms, to some extent, through the deformation of cartilage, but under high loads mostly by local deformation of the subchondral cancellous bone. Note what happens to the stress distribution if joint conformation does not occur.

2 STRESS DISTRIBUTION WITHIN JOINTS

The load across the joint is the (vectorial) summation of (1) body weight plus the forces caused by acceleration and deceleration of the segment, and (2) the muscular forces required to stabilize the joint and move the limb. The contribution from the muscular forces in most cases provides the bulk of the overall force across the joint (see Fig. 2-46). To minimize the stress across the joint this load is distributed over a contact area that is substantially greater than that of the bony shafts, creating the familiar bulbous ending of the bones at their articulations.

The bearing surface of joints consists of two thin layers of cartilage separated by an extremely thin layer of synovial fluid. The cartilage sits on two relatively thick pads of cancellous bone. To minimize the stress on the cartilage it is desirable to distribute the load over as large a contact area as possible. As the joint is loaded, the cartilage and bone deform. Although the cartilage is about 10 times more compliant (less stiff) than the underlying cancellous bone, the cartilage is relatively thin and thus its total actual deformation is limited. The cancellous subchondral bone, although stiffer, is of sufficient depth to permit significant total deformation and thus allow the joint to conform maximally under load, creating the largest possible load-bearing contact area (Fig. 4-1).

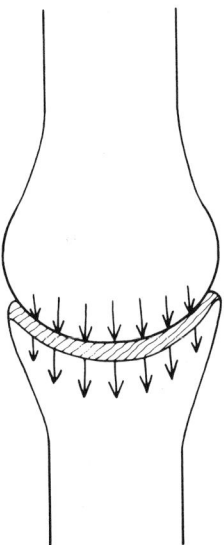

Fig. 4-2. Cartilage acts mainly as a bearing surface, transmitting applied forces to the underlying bone.

Cartilage acts mainly as a bearing surface and transmits the stress applied to it to the underlying bony bed (Fig. 4-2). The subchondral cancellous bone of the metaphysis then acts in two ways: first, by deformation of the bone the joint achieves a maximum contact surface and thus a maximum load-bearing area under high loads; second, the cancellous bone is arranged in trajectories that transmit the major part of the stress down onto the shaft (Fig. 4-3).

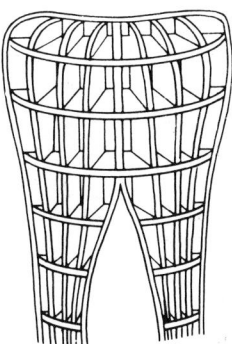

Fig. 4-3. Schematic representation of the cancellous structure of metaphyseal bone. The longitudinal plates of bone act to transmit the joint stress down onto the diaphyseal shaft. These longitudinal plates are reinforced by transverse interconnecting plates or struts.

Trabecular bone deformation also provides some minor shock absorption and thus acts to absorb energy. Trabecular microfracture naturally occurs. This energy of fracture is thus absorbed by the bone tissue. As long as the frequency of these microfractures is low compared with the rate of healing, significant alterations in the deformability of the cancellous bone is not evident. However, because of the importance of subchondral bone in the congruence of joints under load, loss of the compliance of the subchondral bone increases articular stress and can lead to high local stress concentration in the overlying articular cartilage (Fig. 4-1).

3 MECHANICAL BEHAVIOR OF ARTICULAR CARTILAGE: VISCOELASTICITY

Under load, articular cartilage deforms by outflow of water and small solutes that are trapped by the highly hydrophilic cartilage matrix of collagen and proteoglycan. The rate of water outflow is obviously more rapid in the early stages of deformation than in the later stages. As the interstitial "pores" through which this flow takes place become narrower, the material is further squeezed. Since the deformation of a fully soaked sponge, as cartilage, is related to the amount of trapped water that has run out, squeezing with a constant load results in a nonlinear deformation. Initially water comes out easily, and it deforms rapidly. The more the cartilage is wrung out the more difficult it is to get the last bit of water out.

The second unique behavioral characteristic of such a system is that the deformation is closely related to the rate at which the external force is applied. The faster it is squeezed, the harder it is for the water to come out; the more slowly it is squeezed, the easier it is to get all the water out. Such rate dependence of deformation is different from the deformation of common engineering solids. Materials such as wood and metal deform a given amount for a given amount of stress in a linear fashion, elastically. Cartilage deformation is nonlinear as it depends on fluid flow. This strain-rate-dependent deformation is called viscoelasticity. (Bone is viscoelastic as well, but to a lesser degree.) *Viscoelasticity* means that when squeezed quickly, a material will tend to act more stiffly than when squeezed slowly. Injury or damage to a viscoelastic material is therefore more likely to occur at high loading rates (eg, impulsive loads). It is obvious that microdamage will be cumulative if the impulsive loading is repetitive.

The hydrophilic matrix of cartilage tends to maintain water within its substance, creating pressures within it. This pressurized liquid can carry load, as can the pressurized abdominal contents described in Chapter 1, Section 8. As discussed there, the carrying of load by such equilibrated pressurization is called hydrostatic pressure and has a high compressive yield stress.

4 MECHANICAL FACTORS IN THE WEARING AWAY OF ARTICULAR CARTILAGE

Although chemical, enzymatic, and metabolic factors can lower the yield strength of articular cartilage, the wearing away of the bearing surface down to bare bone requires mechanical forces. Termites can weaken a piece of wood, but the wood never breaks up into pieces unless it is subjected to some sort of a loading stress.

From a mechanical point of view, one can separate cartilage fibrillation initiation, propagation, and loss of substance. Cracks and tears beginning in the surface tangential fiber layer must, by definition, be initiated by tensile stress (ie, a pulling apart of the structure). As is discussed in Section 6, joints are lubricated so well that shear forces at the articulating surfaces at best play only a secondary role in cartilage wear. The major loads across the joint are compressive in nature. How then are tensile stresses produced?

If articular cartilage were compressed evenly across its entire surface, no tensile stresses would exist. In whole joints, however, this is probably never the case. Only a part of the joint surface is load bearing at any instant. Since adjacent areas of the cartilage surface are connected, if one area is compressed and another is not, the tissue connecting them is stretched in tension. In this way tensile stresses occur at the periphery of loaded areas (Fig. 4-4).

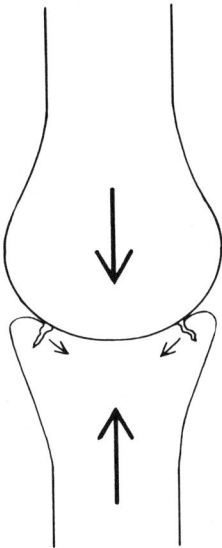

Fig. 4-4. Tensile stresses occur in articular cartilage at the margins of the contact zone. Fibrillation initiates there.

Articular cartilage is designed so that it has maximal resistance to being broken apart. The fiber phase of articular cartilage is collagen. Its general arrangement (unloaded and loaded) is shown in Figure 4-5. At the surface it is arranged to resist tensile stresses by being oriented tangential to the surface. Despite this, it can fail under repetitive normal loads, which accounts for the almost universal findings of fibrillation around the periphery of joints in older persons (see Fig. 4-4).

The propagation of cracks in cartilage depends on the same factors as does their initiation. An increase in the length of a fibrillation (crack) or ultimately even the possible tearing away of an area of cartilage from the calcified bed must involve the generation of tensile stresses. Local stresses within cartilage are dependent on the applied stress, Young's (elastic) modulus of the fibrillated cartilage, Young's modulus of the bony bed, and the stress gradient between adjacent areas of subchondral bone.

The midzone fibers of collagen, although randomly oriented in position in the unloaded state, under compression line up in an optimum configuration along tensile lines to resist crack propagation. If enzymatic degradation or cellular metabolism has weakened this structure, normal levels of repetitive stress could create failure conditions. If this has not occurred, high stress levels must be present to propagate a crack. Such high local

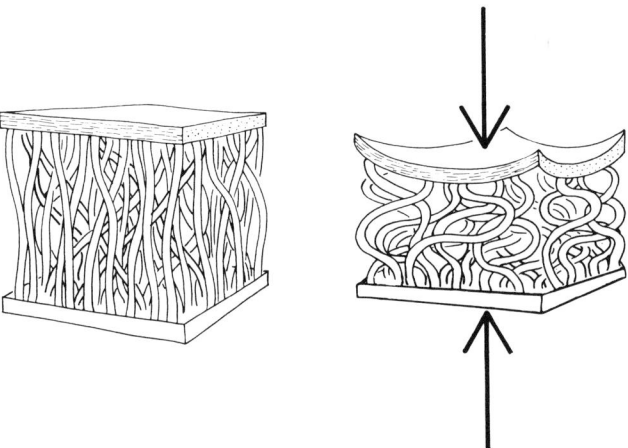

Fig. 4-5. Schematic of cartilage collagen. Unloaded, the orientation of articular cartilage collagen fibers in the central zone is fairly random. Upon compression, the predominant collagen fiber orientation in the middle zone lines up to the load. The precise anatomy of the collagen fiber network connecting the articular cartilage in its calcified bed has been difficult to discern. It has been suggested that the fibers do not regularly span the tidemark between the articular cartilage and its calcified bed.

tensile stress concentrations can occur where the compressive stresses created in the cartilage are uneven. Such a situation arises when cartilage has structural or geometric alterations secondary to inherited or developmental abnormalities or attempts at cartilage repair.

Uneven compressive stresses in the cartilage may not only arise from irregularities within its own structure, but also from uneven stresses in the underlying subchondral bone in continuity with the cartilage. Normal joint structure, even in its deepest layers, is designed to prevent any sharp stress gradients. At the junction between relatively deformable articular cartilage and stiff cancellous bone is an intervening layer of calcified cartilage of intermediate Young's modulus, which assists in the smooth transmission of stresses. Some of the deepest zone of collagen fibers cross these layers and act as an anchor. Furthermore, the junction between cartilage and its calcified bed is not straight but undulates. This increases the surface area and maximizes the possibility of transmitting compressive stresses, rather than tensile stress, from shear.

More significantly, stresses in the deep layer will be highly magnified if the articular cartilage layer becomes thinned. How can articular cartilage wear off? Under what circumstances can the shear stresses on the articular cartilage be severe enough to cause it to fragment and lift off its calcified bed? Although the possibility of the articular surface wearing away exists, because its frictional resistance from surface sliding is so low (coefficient of friction below 0.005), abrasive wear is unlikely. But as the articular cartilage is compressed, it tends to spread over its calcified base. This tendency to spread creates high stresses, particularly at the edges of the contact area, and can promote the development of very deleterious horizontal splits deep in the articular cartilage layer.

Thinning of the articular surface will have a profound effect on increasing the shear stresses acting on the articular cartilage. It has been observed in osteoarthrosis that the articular cartilage layer is frequently thinned. What causes this if not its wearing away? The answer is that the calcified bed the articular cartilage rests on is really the formerly active center for joint growth. It stops being active when we stop growing. It is believed that repetitive impulsive loading causes a reactivation of this dormant growth center with thickening of the calcified plate on which the articular cartilage rests. This thickening is created by an invasion of the calcified front into the articular cartilage (Fig. 4-6). The cartilage thins at the expense of its thickened subchondral calcified layer.

Steep gradients between adjacent sections and transversely across the joint can occur in the deep layer from the natural loading patterns in certain joints (Fig. 4-7). For example, such an area is in the medial central facet of the patella, a common site for cartilage fibrillation. Such local steep stiffness gradients can also occur from alterations in the stiffness of the subchondral

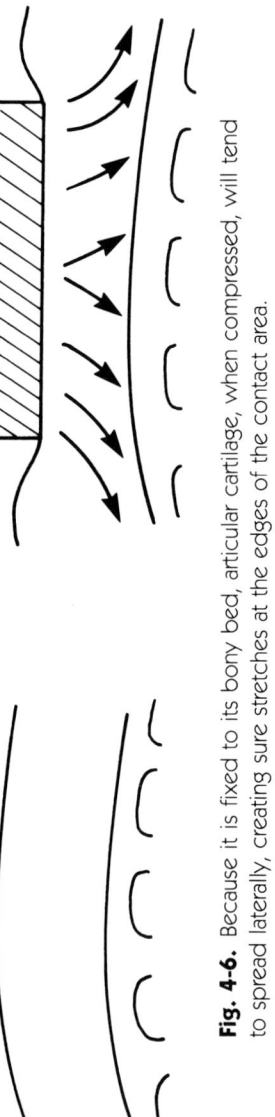

Fig. 4-6. Because it is fixed to its bony bed, articular cartilage, when compressed, will tend to spread laterally, creating sure stretches at the edges of the contact area.

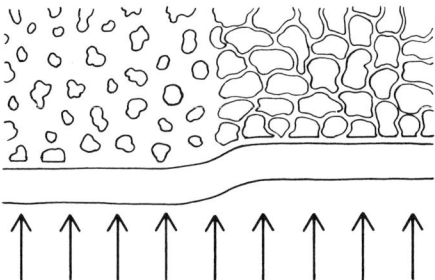

Fig. 4-7. Sharp gradients in the stiffness of the underlying subchondral bone create tensile stress in the cartilage overlying the gradient and can lead to cartilage fibrillation in this area.

scleroses secondary to intra-articular fractures. Multiple repetitive impulsive loads, poorly handled by the musculoskeletal shock absorbers (see Chapter 3) can so stiffen subchondral areas of bone.

Although local stiffness gradients can be associated with cartilage fibrillation, the progression of this fibrillation to total loss of cartilage substance down to bare bone appears to be related to the relative stiffness of its underlying cancellous bone. Osteoporotic, relatively compliant bone appears to spare its overlying cartilage from severe mechanical deterioration, probably because on such a compliant bed sufficient shear stresses in the cartilage depths cannot be generated. (It is easier to abrade tape off a hard surface than off a sponge.) As Newton's first law has taught us, the shear force generated cannot exceed the capacity of the tissues to provide an equal and opposite force.

5 MECHANICAL CONSIDERATIONS IN THE TREATMENT OF OSTEOARTHROSIS

Cartilage wear results when the ultimate tensile strength of cartilage is exceeded. However, contrary to popular teaching, once this process is started, it is not always irreversible. There are well-recorded instances of clinical remission of osteoarthrotic joints, and remission has been achieved experimentally and by surgical intervention. All the metabolic evidence accumulated to date suggests that the joint is capable of a healing response. Failure to heal may be due to a persistent relatively high level of stress in the joint. If these stresses can be sufficiently lowered, one can expect some

144 PRACTICAL BIOMECHANICS FOR THE ORTHOPEDIC SURGEON

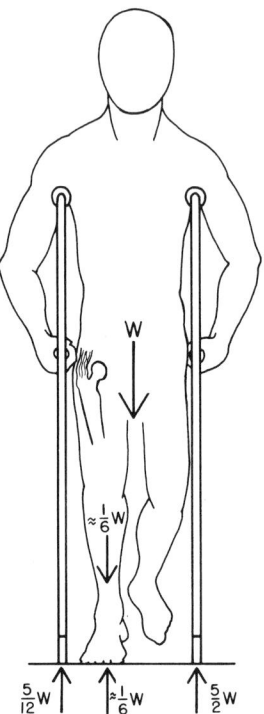

Fig. 4-8. The effect of a touch-down crutch gait is to eliminate almost all of the joint loads in the lower extremity by making it unnecessary for the muscles to stabilize the leg. Since the normal joint forces in the lower extremity are several multiples of body weight (w), the residual weight of the leg (about 1/6w) can almost be ignored.

functional healing of both bone and bearing surface to take place. Stress in a joint can be diminished in one of two ways: the overall load on the joint is decreased or the surface area over which that load acts is increased.

Decreasing the Overall Load on the Joint

The use of external aids for ambulation (walker, crutches, or cane) throughout the world is the most common form of treatment of osteoarthrosis. Of these aids, a walker most disrupts the normal gait cycle. It must be picked up and moved at each step, which interrupts the normal rhythm of gait and denies the use of momentum as an energy-saving device. It is popular with older persons because it significantly enlarges their base of support and does not require significant shoulder and upper upper arm muscles. In actuality, the walker is the external aid of choice for patients with a low energy reserve.

5. MECHANICAL CONSIDERATIONS IN THE TREATMENT OF OSTEOARTHROSIS

Bilateral crutches allow the shoulder muscles to be used effectively and permit some momentum to be applied in the gait cycle. Crutches are most effective when used in combination with a partial weight-bearing gait so that the weight of the limb is rested on the ground, making it unnecessary for the person to support the limb totally in the stance phase (Fig. 4-8). Under such circumstances crutches can free the lower extremity of almost all its load by mainly making it unnecessary to contract the muscles in order to stabilize the hip or leg. Contrary to intuition, a nonweight-bearing crutch gait loads the joints of the lower extremity more than does a touch-down gait, because the muscles must contract to carry the weight of the leg (Fig. 4-9). Theoretically this gait takes about 80% of the load off the joints in the stance phase in that leg compared with 90% to 95% reduction with a touch-down gait.

In a similar manner, a cane obviates the need for the abductors of the hip on the affected side to contract but is much less effective in unloading the leg as muscles still have to contract for some stability (Fig. 4-10). The cane probably relieves about 60% of the load on the hip in the stance phase. It is something on which to lean for support and will have its longest lever

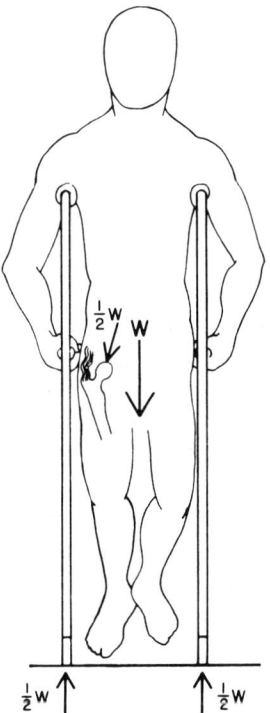

Fig. 4-9. A nonweight-bearing gait requires muscle contraction to balance against body weight. Such a gait is not as load restrictive as a touch-down gait (see Fig. 4-8).

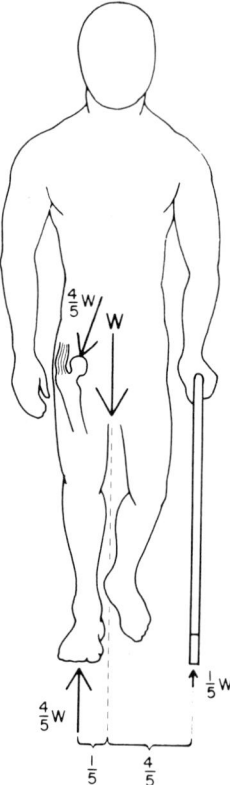

Fig. 4-10. A cane theoretically reduces the load on the opposite leg by about 60%. The amount of load reduction depends on how far away from the body the cane is held.

arm the further away from the affected leg the tip of the cane is put down. The cane improves balance and allows the leg to be brought through with a minimum of joint motion. This can spare inflamed areas of synovium that might be aggravated by such motion.

One can surgically decrease the overall load on the hip joint in two ways. One method consists of lengthening the muscles by tenotomy, such as a Voss or "hanging hip procedure." The length of a muscle and the strength of its contraction are directly related, and when these tenotomized muscles heal with segmental noncontractile elements in series (Fig. 4-11), they are lengthened and their strength is at least one grade weaker than before. A considerable proportion of contraction is dissipated in the scar.

This multiple tenotomy method decreases the overall force, at least transiently, and may provide sufficient time for the joints to heal. Over time the muscles adjust to their new resting length and in some instances regain their initial strength or close to it.

5. MECHANICAL CONSIDERATIONS IN THE TREATMENT OF OSTEOARTHROSIS

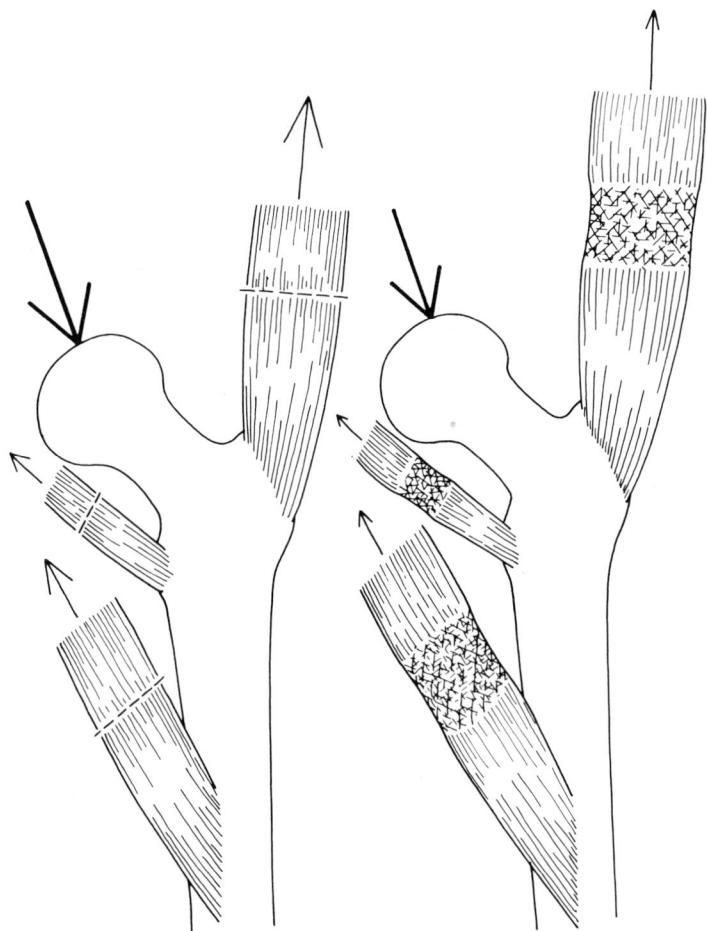

Fig. 4-11. The muscles about the hip can be surgically lengthened, reducing their force of contracture, at least temporarily, by about 20%.

The other method of decreasing the overall load on a joint is by increasing the moment arm of a muscle so that less force is needed to produce a given torque. Around the hip this can be done with a varus osteotomy (Fig. 4-12). In such an osteotomy the lever arm of the femoral neck is lengthened, thus increasing the lever arm of abductors. These balance the forces eccentrically exerted on the hip by the *partial body mass* (ie, the mass of the body minus the leg below the joint being discussed).

As discussed in Chapter 2, the abductors act through a lever arm that is only one-third of the lever arm of the forces exerted by the body mass. Therefore, to ensure moment equilibrium, they must develop a force of about

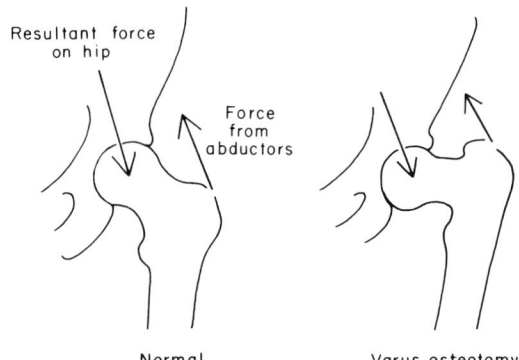

Fig. 4-12. A varus osteotomy increases the lever arm of the abductors and thus decreases the force required by these muscles to produce a given torque.

three times that exerted to the body mass, and the total load supported by the joint approximates about four times the body weight. Lengthening the lever arm of the abductor muscles allows them to achieve equilibrium with a smaller force and hence reduces the overall load on the hip joint (see Fig. 2-46).

The principle of decreasing force by lengthening the lever arm is universal and can be applied to any joint where it is surgically feasible. The compressive force exerted by the patella against the femoral groove can be decreased by lengthening the lever arm of the patellar tendon. This can be achieved by displacing the tibial tubercle anteriorly (Fig. 4-13).

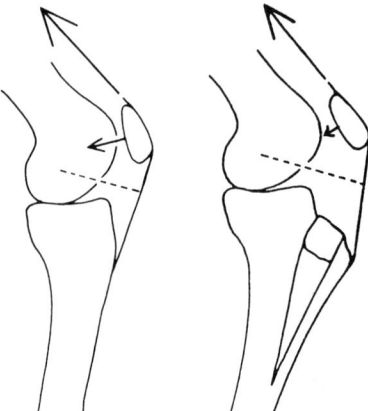

Fig. 4-13. The lever arm of the quadriceps can be increased by anterior displacement of the tibial tubercle.

5. MECHANICAL CONSIDERATIONS IN THE TREATMENT OF OSTEOARTHROSIS

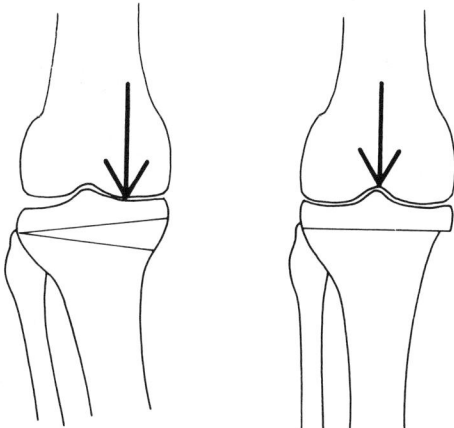

Fig. 4-14. With a varus deformity of the knee most of the load is transmitted through the medial compartment. After valgus osteotomy the load is distributed more evenly.

Increasing the Weight-bearing Area of the Joint

The stress in a joint can be decreased by increasing its load-bearing area. The most obvious example is valgus osteotomy in a varus deformity of the knee. With a varus deformity most of the load is transmitted through just one compartment of the knee. After osteotomy it is evenly distributed (Fig. 4-14). Proper corrective osteotomy allows both medial and lateral compartments to bear equal load. Osteotomy is a well-recognized and accepted measure for treating osteoarthrosis of the knee that is associated with an angular deformity.

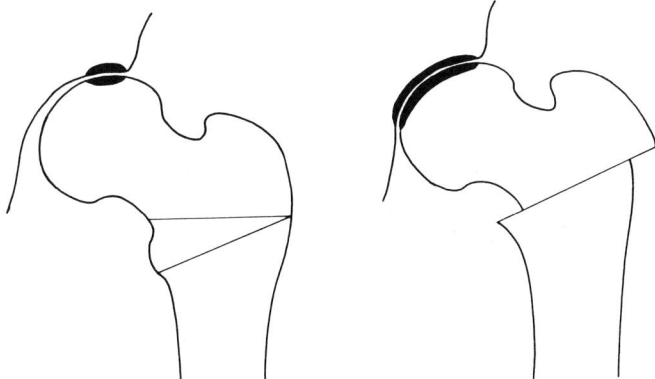

Fig. 4-15. In cases of osteoarthrosis of the hip, where congruity is maintained, varus osteotomy may increase the load-bearing surface of the hip while it decreases the overall load (see Fig. 4-12).

150 PRACTICAL BIOMECHANICS FOR THE ORTHOPEDIC SURGEON

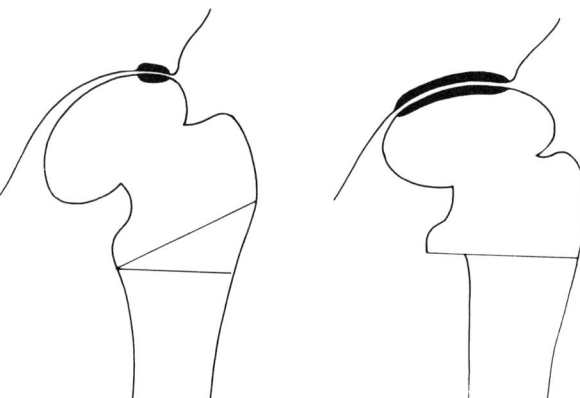

Fig. 4-16. In cases of oteoarthrosis of the hip with a large medial osteophyte, valgus osteotomy can bring that osteophyte into the load-bearing area, significantly increasing the contact area. Since valgus osteotomy increases the force on the hip by decreasing the lever arm of the abductors, concomitant procedures must be done if the stress on the hip is to be significantly decreased.

Osteotomy about the hip increases the load-bearing area of that joint in certain cases. Generally, where congruity is good, especially in abduction of the leg, a varus osteotomy is indicated (Fig. 4-15). In cases where a large medial osteophyte has laterally subluxed the hip (eg, in an old congenital dysplasia) a valgus osteotomy can bring the osteophyte into the load-bearing area, significantly increasing it (Fig. 4-16). Valgus osteotomy does not move the resultant force on the hip lateral to the acetabulum and thus has no tendency to cause further subluxation.

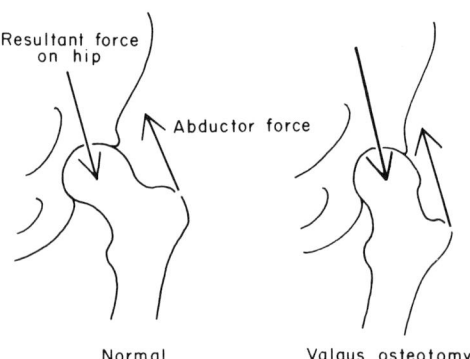

Fig. 4-17. A valgus osteotomy decreases the lever arm through which the abductor functions, increasing the overall load across the normal hip joint.

Valgus osteotomy, however, does decrease the lever arm of the abductors, compelling them to increase their strength of contraction to achieve a torque equilibrium about the hip. This increases the overall load on the hip (Fig. 4-17). For this reason a valgus osteotomy must be carried out in conjunction with a hanging hip or Voss procedure; otherwise, the force on the joint is not lowered, even though its load-bearing surface is increased.

Mechanical Stimulation of Fibrocartilaginous Healing

Fibrocartilaginous healing of areas of destroyed articular cartilage requires a source of cells (usually present either in subchondral bone or in surrounding soft tissue), decreased stress and motion. Active motion or active assisted motion contracts muscles across the joint and increases the stress. Therefore, for the first several months after osteotomy, active exercises should be kept to a minimum.

Experiments have shown that primitive pleomorphic fibroblasts, subjected to equal pressure from all directions, produce fibrocartilage (see Fig. 2-27). Continued motion of the healing joint provokes the formation of synovial fluid and creates stresses in the healing mantle over the articular surfaces. If these surfaces are relatively congruent, and the stresses are not excessive, the pressurized synovial fluid transmits hydrostatic pressure. This tends to promote differentiation of chondroid material. Motion apparently orients the superficial fibers of the healing fibrocartilage tangential to the surface and those fibers next to the bone perpendicular to the surface. Over time, maturation of the chondroid bearing surfaces gives a hyaline appearance to the tissue. Too high a stress in the initial healing periods probably increases the tensile stresses to such an extent that the healing surface tissue is ripped apart.

The healing of full-thickness ulcerations of articular cartilage depends on the diameter of the ulcer. Depending on the curvature of the joint, the shoulders of the ulcer need to be close enough together to keep the opposing articular surface from abrading away the early healing tissue (granulations) that will form in the base of the healing ulcer (Fig. 4-18). Because of this mechanical circumstance, it should be expected that large ulcerations of articular surfaces will not have a chance to heal. Even if large, ulcerations in the knee that articulate with the meniscus have a chance to heal because the meniscus protects the healing articular surface from abrasion of the early healing tissues. It has been suggested that cartilage will heal in joints where the stress is well distributed. In certain circumstances this appears to be the case.

The unreliability of interposition arthroplasty of the hip probably relates to the fact that the surgery and postoperative course do not significantly lower stress for a sufficient period of time. Some also question whether such an extensive procedure creates osteonecrosis under the metal mold, but most

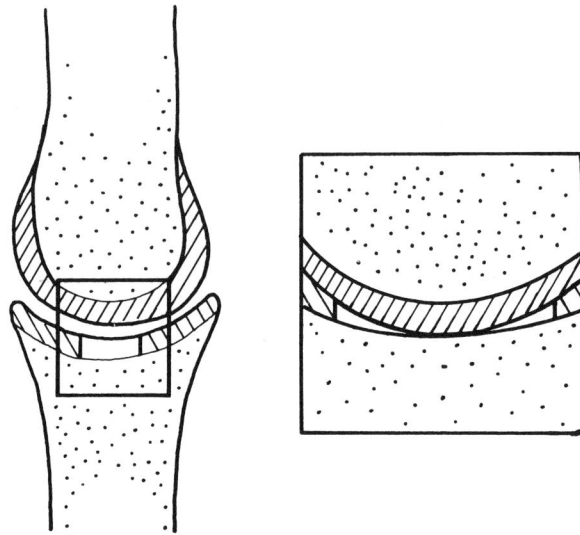

Fig. 4-18. In a large chondral defect, the opposing articular surface will tend to abrade healing tissue in the depth of the defect. The shoulders of a smaller defect will protect the healing tissues.

carefully examined specimens suggest the failures were due to stress concentration rather than to primary vascular insufficiency.

6 FRICTION ACROSS JOINTS

Joint lubrication is discussed at the end of this chapter because failure of lubrication mechanisms is not believed to be primarily responsible for joint degeneration in most clinical situations. Nevertheless, the subject of joint lubrication is important for a complete understanding of joint mechanics and has significant implications in the functioning of total joint replacements, the subject of Chapter 5.

Joints function as the articulations of the skeletal system and must provide relative motion of the body segments while transmitting load. Theoretically, then, during articulation, shear forces arise at the interface of the articulating surfaces, and these shear forces have to be overcome by muscle action to produce motion. Synovial joints undergo a variety of loading and motion conditions during normal activity. These range from relatively low to high loads and with a varying velocity of sliding of the opposing cartilage surfaces. Loading can be intermittent and is often impulsive. Therefore, joints must be lubricated by a mechanism that functions effectively for all types of conditions or by several different lubricating mechanisms in order to ensure low friction under the full range of operating conditions.

The frictional resistance developed between the articular cartilage surfaces in a joint depends on how much the surfaces actually touch. The less they touch, the easier the two surfaces slide. Obviously the amount of force tending to drive one surface into another (ie, the load applied across the bearing) affects how easy it is to push or pull one surface over the other.

The constant that relates load and frictional resistance is the *coefficient of friction*. It is obtained by dividing the load (kiloponds or pounds) into the force of frictional resistance (kiloponds or pounds). Thus the coefficient of friction is a unitless measure by which the frictional resistance of various bearings can be compared. It is independent of the amount of surface areas in contact with each other. Examples of approximate coefficients of friction are a steel-on-steel bearing lubricated by oil, 0.210; a plastic-on-metal total hip replacement lubricated by synovial fluid, 0.060; an ice skate on ice lubricated by water, 0.030; and articular cartilage on articular cartilage lubricated by synovial fluid, 0.002. These numbers show that a synovial joint is 100 times easier to move than a steel-on-steel bearing and about 30 times easier to move than a plastic-on-metal total hip replacement.

Note that in the above examples the lubricant is always specified. Lubricants act to keep the surfaces apart, thus decreasing the shear forces acting between them. A well-oiled bearing moves more easily than a dry one. As the friction developed is the resistance to shear, in most cases it is much easier to shear a fluid lubricant than a solid surface. One property of a well-lubricated system is that under the conditions of load that are applied the lubricant is always present and is not squeezed out of the bearing interface or else its effect is diminished.

Two major mechanisms function at the cartilage interface to maintain lubrication, either or both acting according to the load and motion conditions in the joint at a particular time. At one extreme is a *boundary lubrication* phenomenon. Here molecules from the synovial fluid attach themselves by chemical interaction to the articular surfaces.* These bound molecules create a boundary layer that, when rubbed against itself, offers less resistance to shear forces than would rubbing the two bare articular surfaces against each other (Fig. 4-19). This situation can be likened to a wax coating on the floor or to teflon on a frying pan and is one of the many reasons why such surfaces are slippery. In joints, the boundary layer does not operate at high loads as it is too fragile to withstand the shear forces created under such conditions.

How then can a lubricant be kept in an oscillating bearing under high loads and speeds? Several possibilities exist. In most bearings the lubricant is held in place by the physical force generated by the relative sliding motion

---

* The boundary-lubricating molecules for synovial fluid are a moderate-sized glycoprotein, chemically distinct from hyaluronate. The role of hyaluronate in joint lubrication is discussed later in this section.

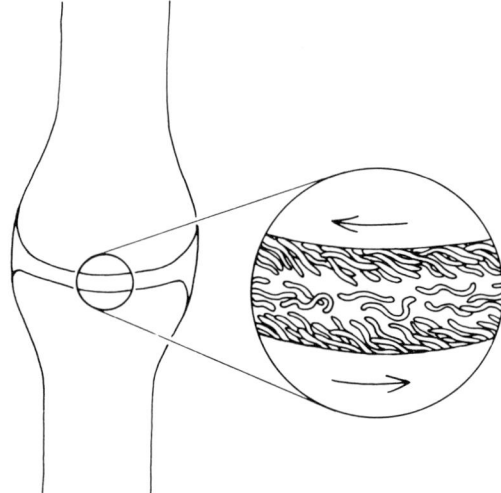

Fig. 4-19. Boundary lubrication involves the binding of lubricating molecules to the bearing surface.

of the bearing pushing the lubricant ahead. This mechanism is called *hydrodynamic lubrication* and depends on the relative motion of the opposing surfaces to maintain the wedge of fluid lubricant (Fig. 4-20).

Hydrodynamic lubrication is unsuited for diarthrodial joints because a lubricating wedge of fluid cannot be maintained in an oscillating bearing. In such bearings, as a wedge of fluid is built up it is quickly destroyed by the rapid reversal in direction of motion of the bearings.

In joints, the lubricating mechanism that appears to be present under high loads and speeds is based on a self-pressurized hydrostatic phenomenon called *weeping lubrication*. Instead of the fluid being pushed forward into the contacting space from behind, it is pushed up from within the substance of the contacting surfaces themselves (Fig. 4-21). As the joint is loaded the lubricant in the zone of potential contact is pressurized, and with cartilage deformation the interstitial fluid from the cartilage is squeezed out of the cartilage around the periphery of impending contact area.*

As the joint slides, the wept fluid is caught between the lubricating parts and adds a low-viscosity component to the synovial fluid already trapped in the area, aiding in the separation of the cartilage surfaces. Pressurized fluid or hydrostatic lubrication is an effective way of maintaining fluid and

---

* The squeezing together of the two bearing surfaces creates pressurization of any fluid lubricant caught between the surfaces. The resulting fluid film of lubricant is called a squeeze film.

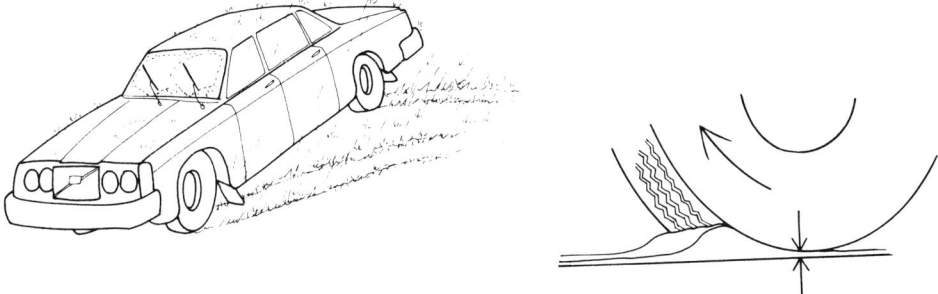

Fig. 4-20. Hydrodynamic lubrication involves the maintenance of a fluid between the bearing surfaces because of the relative motion of the bearing. A wedge of lubricant forms at the leading edge of bearing contact. An automobile skids on a wet road due to loss of friction caused by a hydrodynamic phenomenon.

position against an external force. In industry, pumps are used to maintain hydrostatic pressure. In joints, the articular cartilage under load self-generates pressure that squeezes the fluid in the cartilage out.

Thus the relative sliding motion of the joint accompanied by compression of the cartilage results in the pressurized fluid film between the surfaces composed of entrapped synovial fluid and wept cartilage interstitial fluid (Fig. 4-22). This self-pressurized *hydrostatic lubrication* is most effective in high loads and particularly during transient or impulsive loads.

The relative sliding motion of two compliant surfaces over each other allows some deformation horizontally and may aid in lowering the frictional resistance (Fig. 4-23). This may be considered an additional aid to lubrication and is referred to as an *elastohydrodynamic phenomenon*. It was initially described when rubber coatings were put on steel rollers. Elastohy-

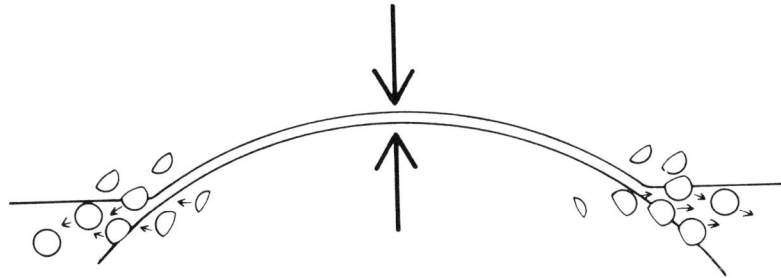

Fig. 4-21. Weeping lubrication results when fluid is pushed up into the joint space at the periphery of the zone of impending contact.

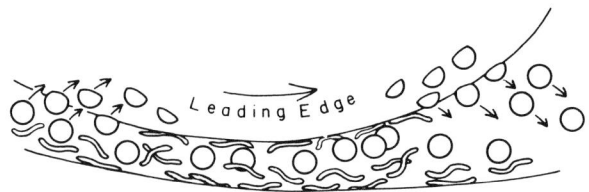

Fig. 4-22. In joints under load the fluid film between the cartilage surfaces is composed of wept interstitial fluid and trapped synovial fluid. Under most physiologic circumstances the fluid weeps out at the leading edge of contact. Osmotic pressure pulls the interstitial fluid back into the cartilage substance in unloaded areas, usually at the trailing edge.

drodynamic effects help to lower the coefficient of friction in articular joints, and the *elasticity* of the bearing surface helps the surface bunch up at the edges to keep the fluid within the zone of impending contact.

During a normal cycle of joint motion, one would expect the prevalent lubricating mechanism to change from predominantly boundary to predominantly weeping and back to predominantly boundary with a combination of both mechanisms active during most of the load and motion cycle.

The effectiveness of this range of lubricating mechanisms over the range of loads and motion encountered in normal joints is demonstrated in Figure 4-24, which compares the friction—a direct indicator of lubricating effectiveness—of the synovial joint under different loads. Note in Figure 4-24 that synovial tissue rubbing over articular cartilage or synovial tissue rubbing on itself has a frictional resistance considerably higher than that encountered in cartilage-on-cartilage lubrication. This is because synovial tissue does not permit the full range of lubricating mechanisms that occur when cartilage is lubricated with synovial fluid. Not enough pressure is exerted to allow for a hydrostatic phenomenon, and the relatively slow velocity of sliding as well as its oscillatory nature precludes a hydrodynamic phenomenon. Synovial tissue rubbing on itself is primarily lubricated by synovial fluid hyaluronate, which adheres to the synovial tissue and provides a boundary type of lubrication.

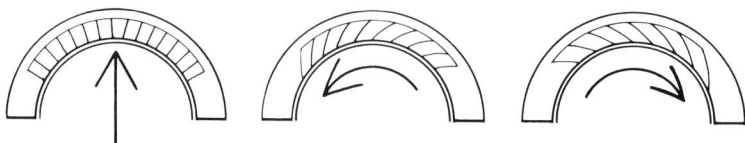

Fig. 4-23. The elasticity of the surfaces helps to lower the functional resistance. This is an elastohydrodynamic phenomenon.

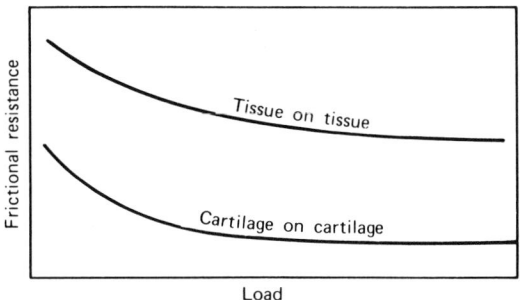

Fig. 4-24. Friction as a function of load for cartilage on cartilage and for synovial tissue on synovial tissue.

The presence of this large molecule, hyaluronate, whose major function is the boundary lubrication of soft tissue, gives synovial fluid significant viscous resistance to flow. The shear forces created by these large, ungainly molecules is significant in that they entangle and entrap each other as the fluid flows (Fig. 4-25). It takes a certain amount of shear to move these molecules along, and this gives the fluid its viscous properties. This entrapment of molecules is obviously less if the fluid flows quickly and is what is meant by *thixotropy*.

Ketchup is a common thixotropic fluid. One pounds on the ketchup bottle to increase the velocity of the flow. Although a thixotropic lubricant, particularly at low shear rates, adds to the frictional resistance of joints, this

Fig. 4-25. The viscous nature of synovial fluid is the result of the entanglement of hyaluronate molecules.

incremental increase is well worth accepting because the viscous nature of the lubricant provides it with significant spreading powers. This is particularly the case when one is concerned with lubrication of soft tissue, which because of its many folds and convolutions might otherwise frequently be without a fluid film or a means of obtaining a lubricant during a particular attitude or function of a joint if the synovial fluid flowed about too freely. The viscous nature of the synovial fluid maintains it spread around in all recesses of the synovium.

7 ATTEMPTS TO TREAT OSTEOARTHROSIS WITH ARTIFICIAL LUBRICANTS

An understanding of the normal lubricating mechanisms shows that joint lubrication depends on the presence of cartilage for both the hydrostatic and the boundary-lubricating mechanisms to function. Particularly under high loads, the hydrostatic (weeping) mechanism is the crucial friction-lowering device. These facts suggest that artificial lubricants, in the presence of mildly destroyed cartilage, are of little use. The addition of oil or silicone blocks the free flow of water in and out of cartilage, effectively destroying the important hydrostatic mechanism.

Attempts to add appropriate boundary-lubricating molecules have thus far been thwarted by an inability to produce molecules that are both hydrophilic and not immediately cleared from the synovial space. The concept that what arthritic joints need is a "good grease job" misrepresents the true situation. The osteoarthrotic joint needs a functioning articular surface.

5

Joint Replacements

1 GENERAL REMARKS

Replacement of the worn out or damaged hip joint by a mechanical bearing has been remarkably successful. More than 90% of the patients who have hip or knee joints replaced have an initial good or excellent result. For some elderly patients, with little functional demand, the result continues to be satisfactory for as long as 20 years. In young patients and in those who have a higher functional level, the initial good result can often deteriorate with time. This leaves the patient and orthopedic surgeon with the difficult task of reconstructing a functional joint. In most cases today, bone and cartilage surfaces are replaced by a bearing couple of polished metal and high molecular weight polyethylene. Joint replacement markedly changes the magnitude and direction of the stresses and the resulting strains in living tissue that can lead to failure of the materials or to an extreme biologic response.

Chapter 4 described how the normal joint functions with an extremely low frictional resistance while maintaining a maximal contact area under load. Joint function is truly a marvel of biology that responds to the mechanical environment by the addition or subtraction of structural elements to keep the deformation of tissues in an appropriate range. Trabeculae are positioned and sized to resist applied stress through the joint. Shear stress is minimized by the astoundingly efficient lubrication of the articular surface. The adaptation of this system to the new mechanical environment of the artificial replacement determines whether an arthroplasty will endure.

Assuming a patient will attempt to return to near-normal function, the total joint replacement is subjected to the same external forces caused by body weight and accelerations of gravity and muscle force as a natural joint. The arthroplasty components themselves are only minimally compliant and transmit largely unaltered forces to the interface with living tissue. The usual shock-absorbing features of cancellous metaphyseal bone in normal joints is based on fine tuning of the bone structured by remodeling. This metaphyseal cancellous bone is damaged and frequently removed at arthroplasty, and a different bed of cancellous or endosteal bone is exposed.

160 PRACTICAL BIOMECHANICS FOR THE ORTHOPEDIC SURGEON

This bone, now thrown into a position of support for the arthroplasty components, must adapt to the new mechanical environment through a reorganization of structure to meet the new demands.

The prosthetic components must be anchored to the skeleton. It is generally agreed that the more rigid the attachment the less painful the end result will be. Rigid fixation of a rigid device to living bone (which requires some deformation for healthy remodeling) is a mechanical problem that has not been fully solved. The most successful fixation to date is still that achieved by methylmethacrylate bone cement. This material acts as grouting by filling the space between the prosthesis and the supporting bone,

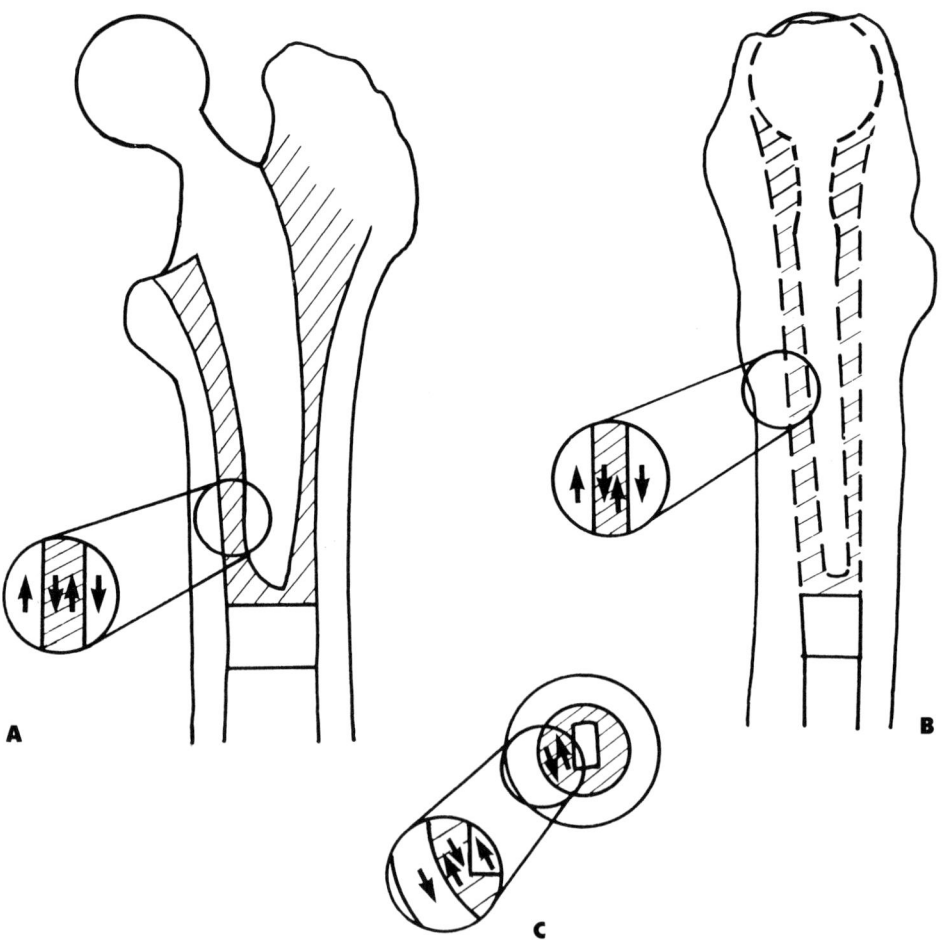

Fig. 5-1. Shear stresses are transmitted from the prosthesis shank to the endosteum at the cement-prosthesis interface and cement-bone interface as demonstrated in the **(A)** coronal view, **(B)** sagittal view, and **(C)** transverse view.

interdigitating with the trabecular or endosteal structure of bone to provide fixation (Fig. 5-1). (Bone cement is not an adhesive material.) Because of its pivotal role in successful fixation, bone cement has been the subject of intense investigation to improve its mechanical properties.

With passage of time, biologic changes at the interface between nonliving material and bone can lead to the formation of a layer of fibrous tissue. This fibrous layer's mechanical properties change the stresses and strains in the surrounding bone and prosthesis. Because this fibrous tissue layer is present in many, but not all, patients and because its properties are not fully characterized and probably change with time, the goal of a complete understanding of the biomechanics of the replaced joint has not yet been attained. Still, understanding the mechanics of the joint to be replaced and the biologic response as evidenced by the clinical functioning of a replacement is helpful in understanding total joint arthroplasty.

To explain the biomechanics of joint replacement we have elected to discuss the two most frequently replaced joints, the hip and the knee.

2 FORCES IN THE NORMAL HIP JOINT

In a normal joint, stress distribution depends on the magnitude and direction of resultant force transmitted through the joint. Shear forces (ie, forces parallel to the surface) are negligibly small in normal joints because of the extremely low coefficient of friction.

The resultant force through the hip can be calculated for two-leg or single-leg standing, for normal gait, or for other specific functional activities such as climbing stairs or rising from a chair. The calculated resultant force (confirmed by experimental observations with instrumented prostheses) depends on the activity and rate at which the femur is accelerating the body either against gravity alone or combined with inertial or frictional effects in the performance of that activity (Fig. 5-2).

In simple walking, forces several times body weight, directed from medial to lateral by approximately 15° from the vertical and alternating anterior to posterior and posterior to anterior, have been calculated. This force subjects the acetabulum and femoral head to compressive stresses. Analysis of the forces about the hip have classically been limited to the coronal plane. In single-leg stance, such an analysis is a good approximation, but functionally it is not. Discussion of the forces on the hip will assume a "functional" joint with forces present in all three planes.

On the acetabular side, the magnitude of compressive stress decreases as it radiates from the joint surface in the bone of the pelvis (Fig. 5-3). On the femoral side, stresses increase as the larger area of the femoral head gives way to the smaller diameter of the femoral neck and the calcar femorale region (Fig. 5-4). A review of radiographs of the hip joint reveals that the bone of the hip joint, fine tuned by remodeling in response to the loads,

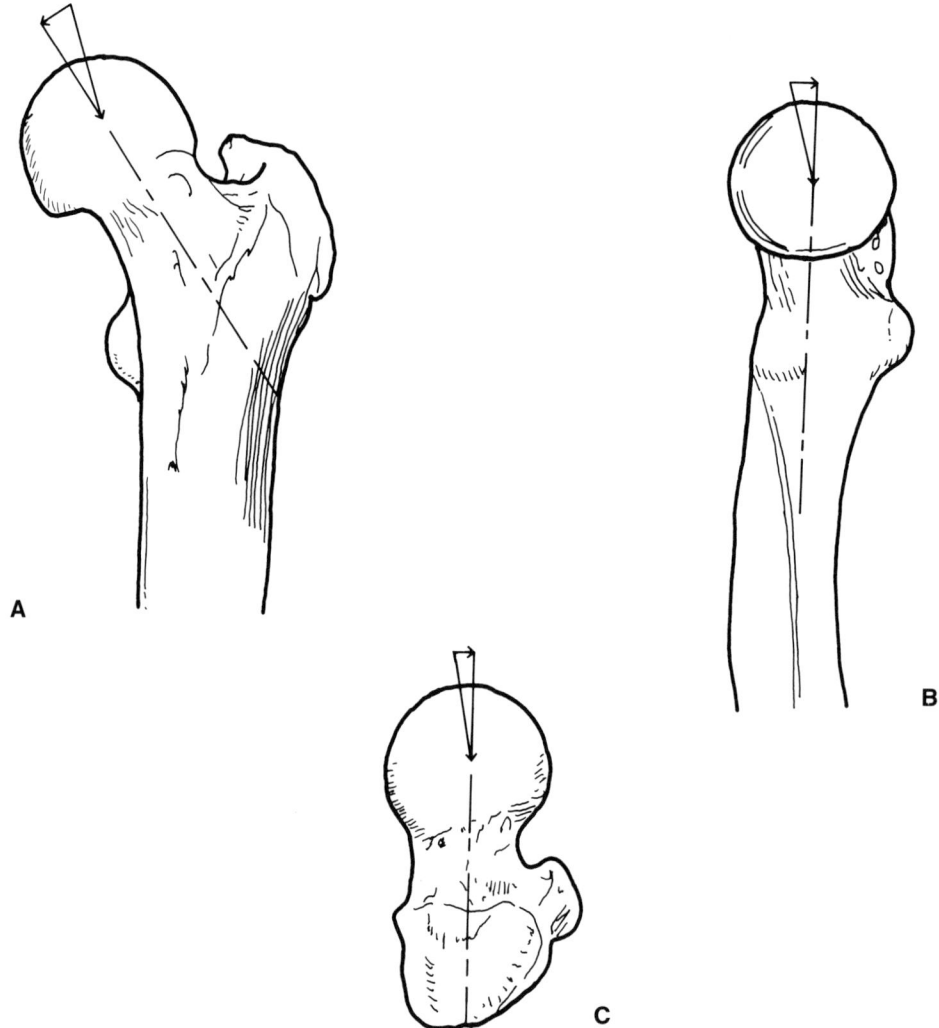

Fig. 5-2. Loads are directed laterally and posteriorly, leading to deformations as demonstrated in the **(A)** frontal, **(B)** sagittal, and **(C)** transverse planes.

shows condensation in areas of predicted high stress. The bone can be thought of as attempting to adjust its mass to preserve some predetermined strain magnitude.

Distribution and magnitude of stresses are unique in the femur because the resultant force acting on the femoral head is not parallel to the axis of the neck (Fig. 5-5). The force acting on the head creates a bending moment

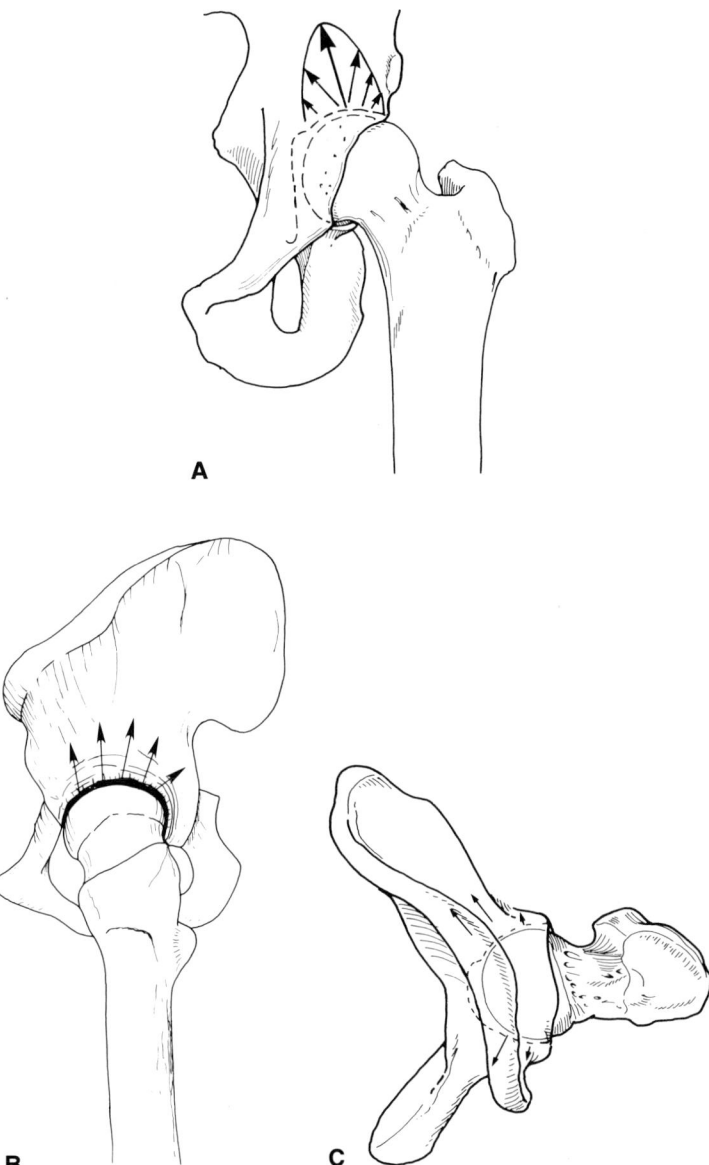

Fig. 5-3. In the normal hip joint, stresses transmitted from the femoral head to the acetabulum radiate out and decrease in magnitude as they are transmitted into the pelvis. **(A)** Coronal view. **(B)** Sagittal view. **(C)** Transverse view.

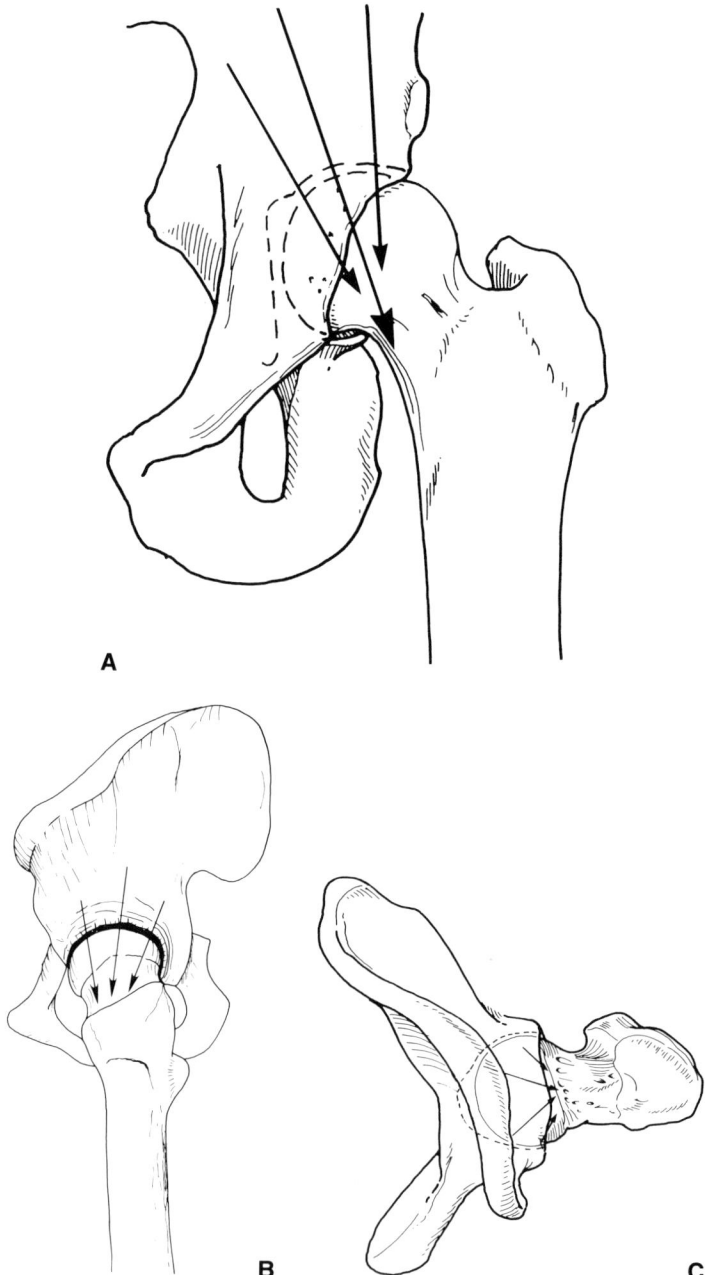

Fig. 5-4. The stress distribution from the total resultant force on the normal femoral head is compressive and well distributed over the femoral head, but becomes concentrated in the calcar. **(A)** Coronal view. **(B)** Sagittal view. **(C)** Transverse view.

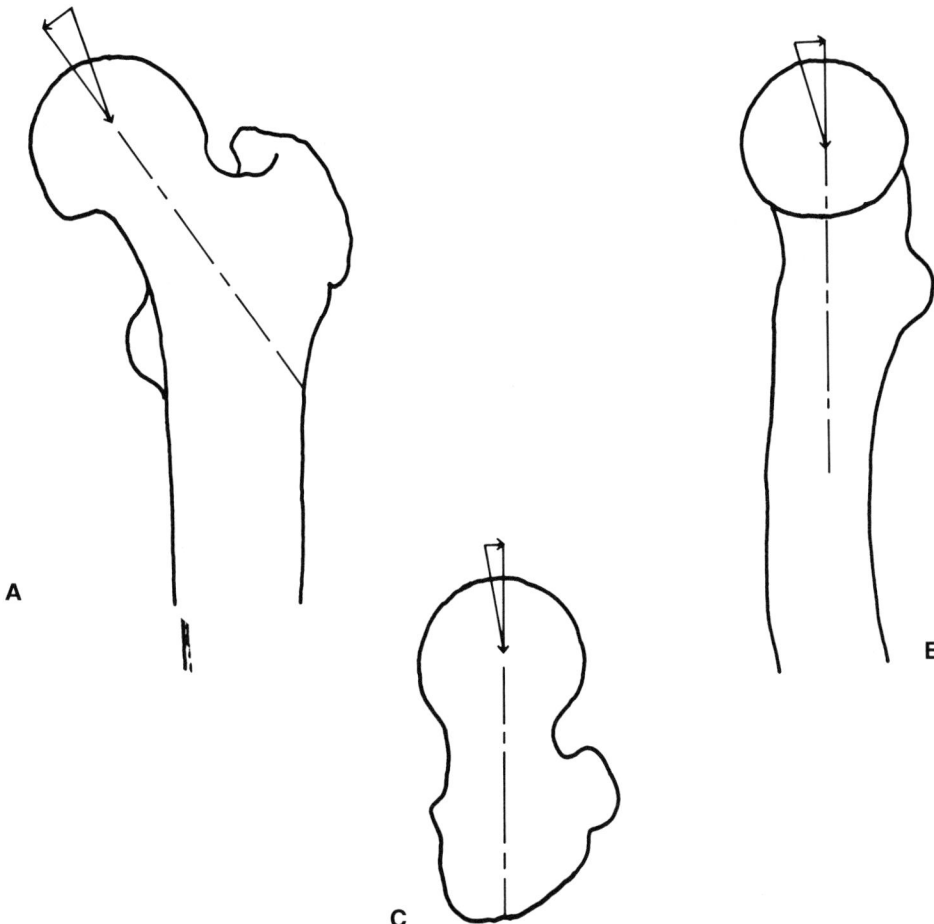

Fig. 5-5. The resultant force on the femoral head has components along and perpendicular to the femoral neck axis. **(A)** Coronal view. **(B)** Sagittal view. **(C)** Transverse view.

leading to compressive stresses on the medial aspect of the neck and to tension stresses on the anterior aspect. These stresses increase in magnitude as they progress from the joint toward the base of the neck as the bending moment is increased (Fig. 5-6).

Note that the entire neck is subjected to some compressive stress since a component of the resultant force is still directed along the axis of the neck (Fig. 5-7). The compressive stress adds to the compressive stress from bending stresses in the medial aspect and decreases the tensile stresses in the lateral aspect (Fig. 5-8).

Stresses caused by bending are greater in magnitude than are the stresses

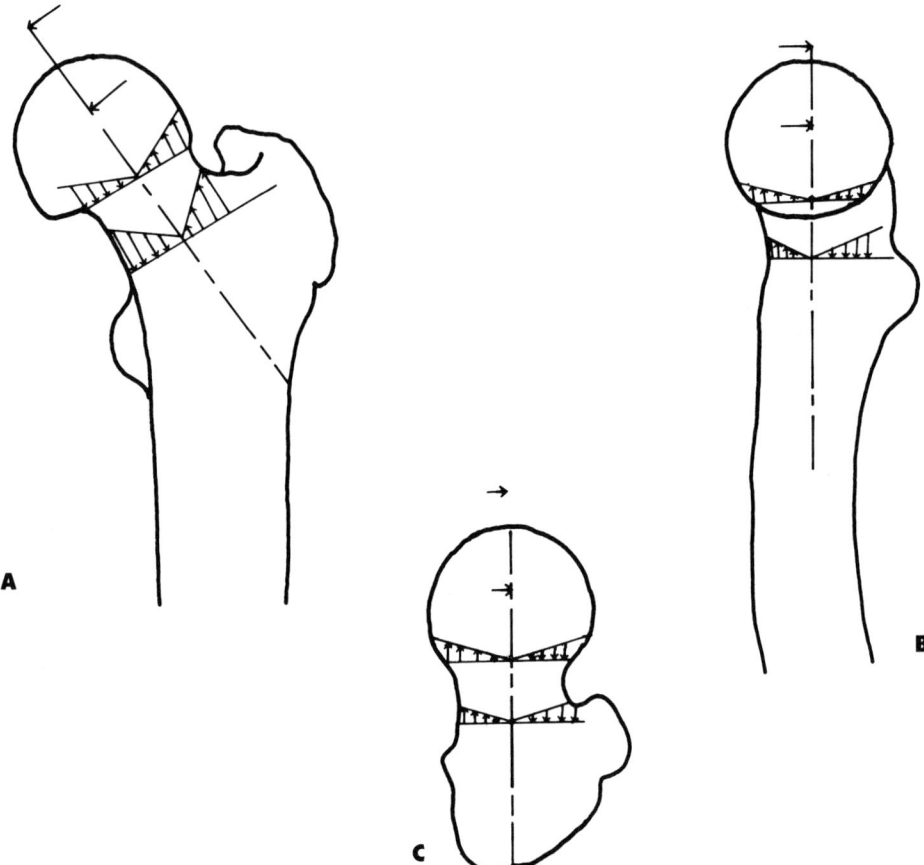

Fig. 5-6. Bending stresses on the femoral neck in the **(A)** coronal, **(B)** sagittal, and **(C)** transverse plane for situations where the resultant force is directed posteriorly. Note compression medially and posteriorly and tension laterally and anteriorly.

from pure compression. If the femoral neck is directed in a line more parallel to the resultant compressive force (ie, more valgus position) the bending would be decreased or even eliminated. In such a situation, only the relatively small stresses from compression would remain. If the femoral neck is directed more transverse to the resultant force, the bending moment would be correspondingly increased while the compressive force would remain almost constant. Keep in mind that these changes in stresses and stress patterns take place without change in the resultant force, but rather from a change in the angle of the femoral neck.

Abductor muscles produce a force on the trochanteric region that acts in a nearly vertical direction in all three planes to balance the force from body

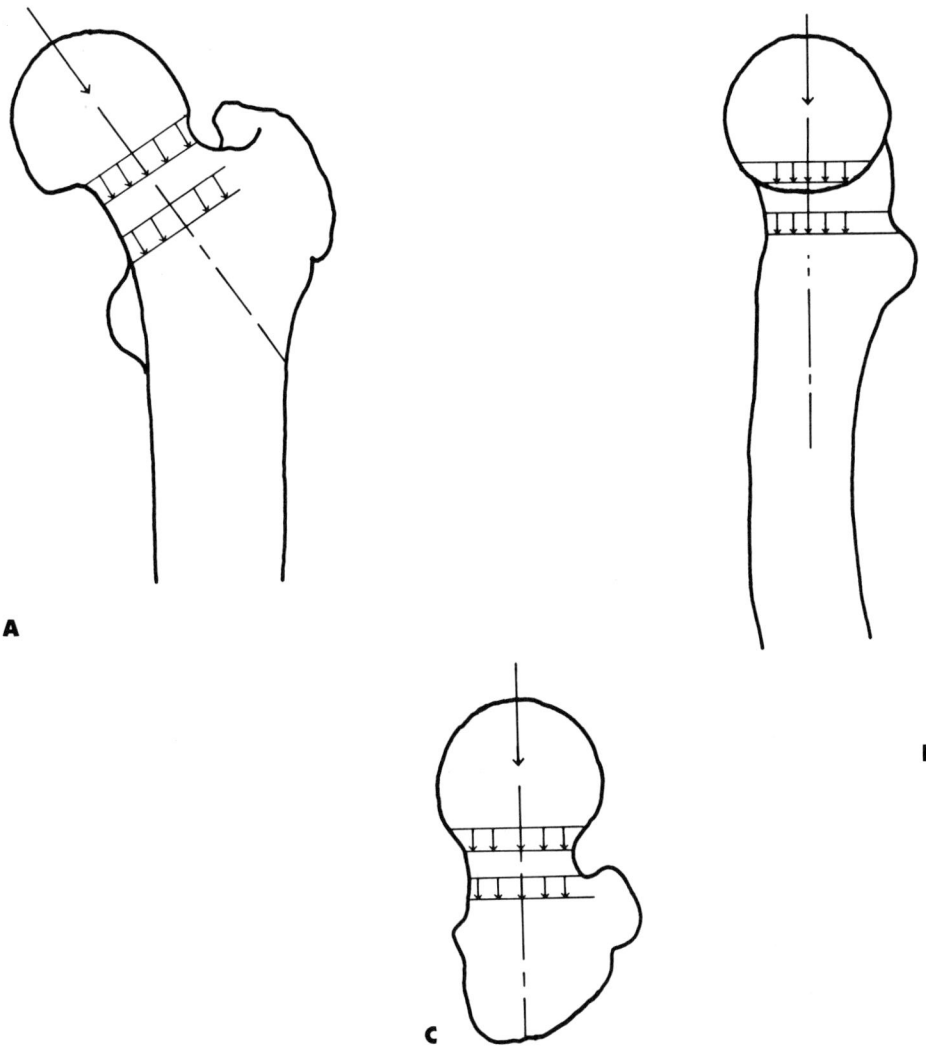

Fig. 5-7. Stress in the femoral neck due to the axial component of the resultant force is compressive regardless of anterior to posterior or posterior to anterior direction of the resultant force. **(A)** Coronal view. **(B)** Sagittal view. **(C)** Transverse view.

weight in that plane (Fig. 5-9). The femoral neck, which lies at a distance from application of force by the abductors, acts as a lever arm for action of these muscles. Abductor muscles thus produce their own bending moment on the neck, increasing tensile stresses medially and compressive stresses laterally. Thus stress on the femoral neck, as is true for any part of the femur or for any bone in the skeleton, results from the force exerted by the

168 PRACTICAL BIOMECHANICS FOR THE ORTHOPEDIC SURGEON

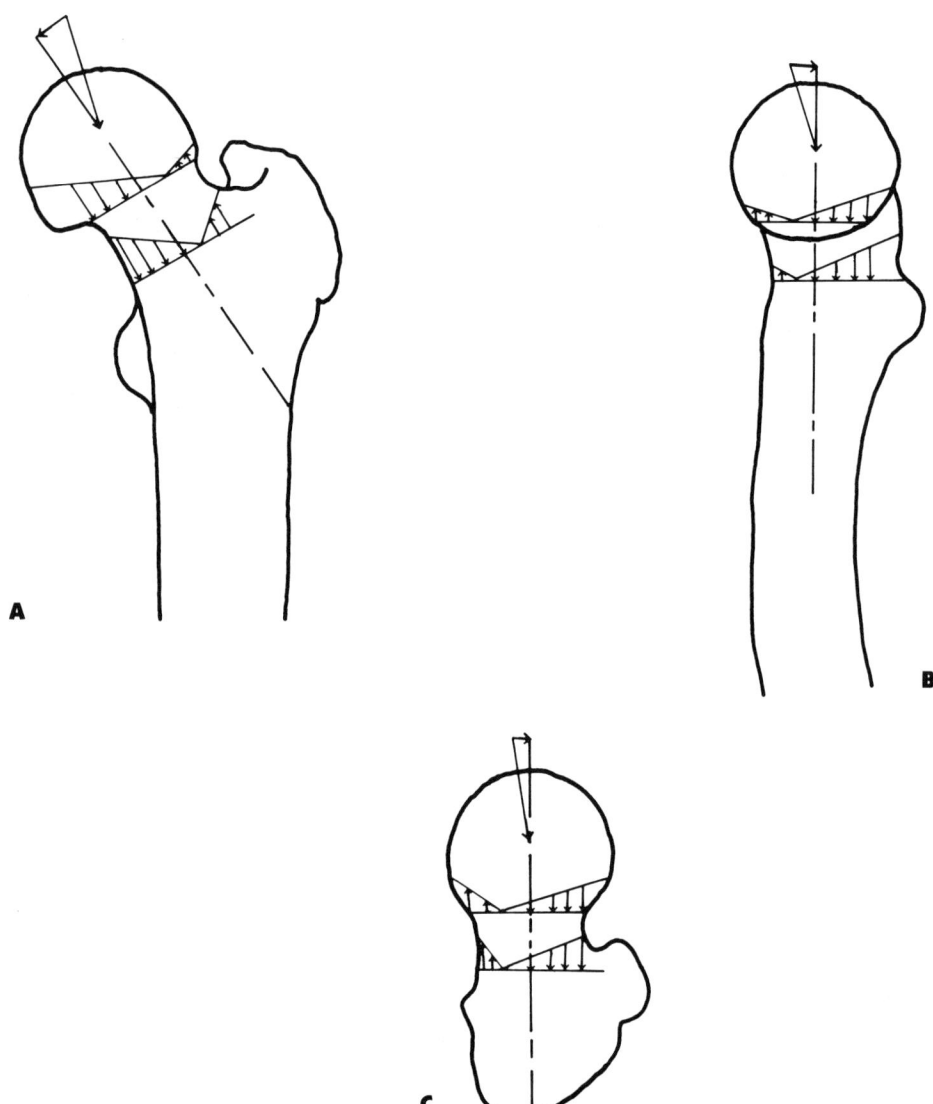

Fig. 5-8. The combination of bending and compressive stress in the femoral neck acts to decrease the tensile stress. **(A)** Coronal view. **(B)** Sagittal view. **(C)** Transverse view.

partial body mass supported by the bone added to the force of the muscles acting on the bone through contraction. This stress depends on the activity being performed and on the rate of acceleration in that activity.

The calculated resultant stress for single-leg stance (a static situation) has been shown to produce bending in a plane close to the coronal plane.

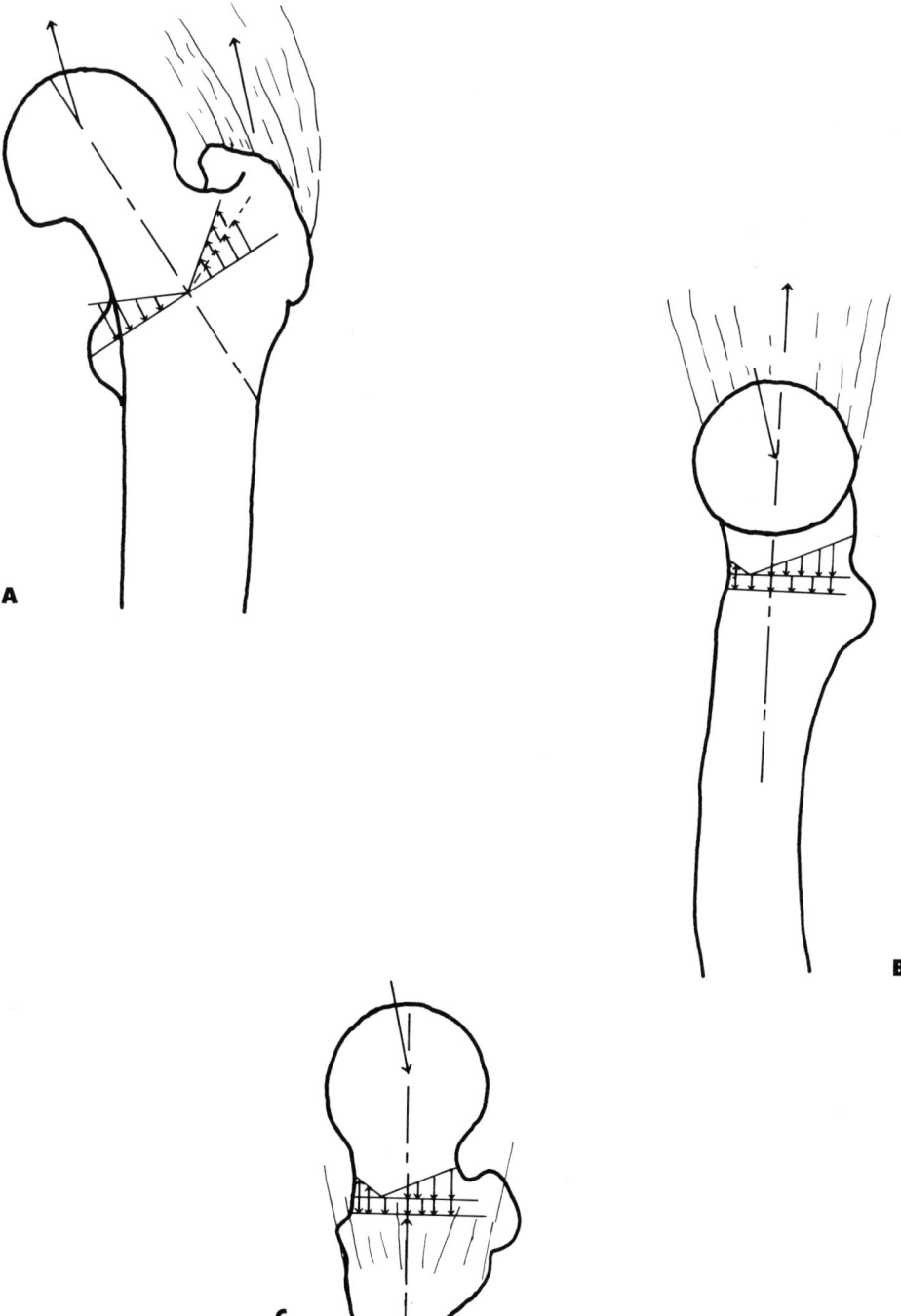

Fig. 5-9. The force created by the abductor tends to increase compression laterally and tension medially in the **(A)** coronal plane. With the resultant force from the abductors along the axis of the neck, the abductors increase compression in the **(B)** sagittal and **(C)** transverse planes.

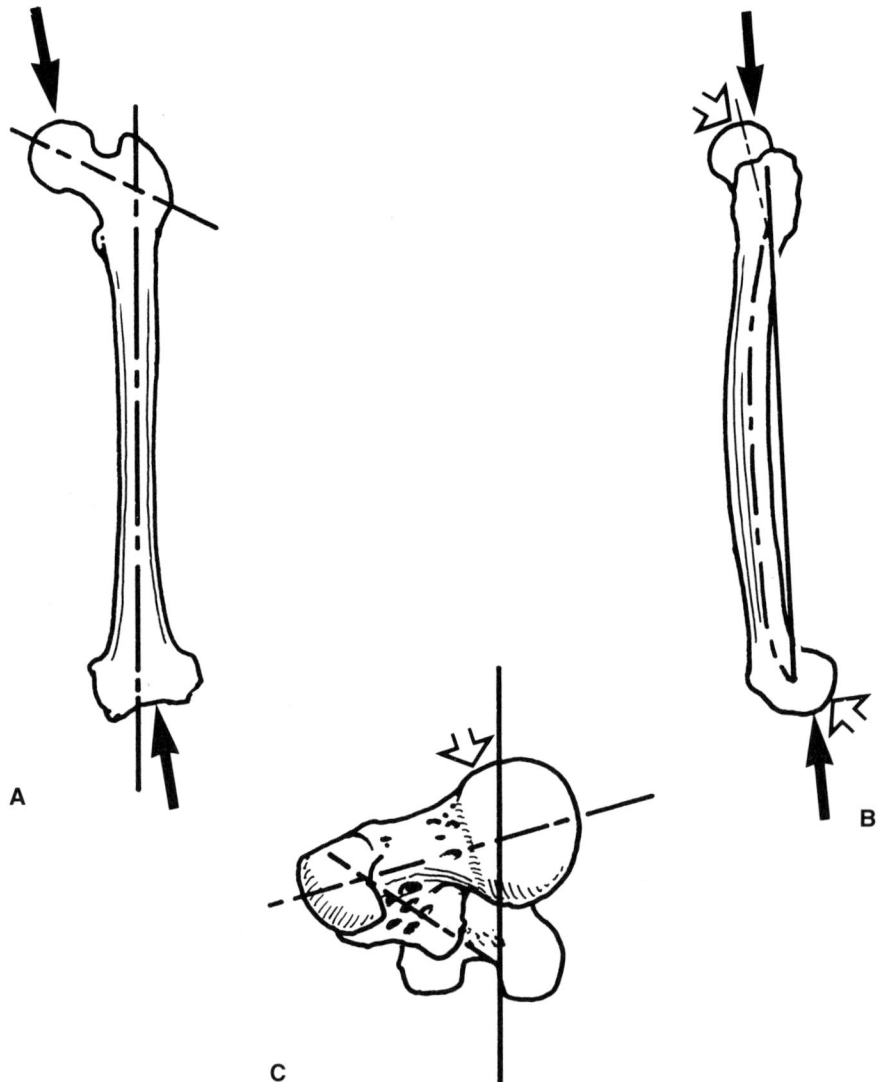

Fig. 5-10. The forces on the femur tend to bend and twist it. **(A)** Coronal view. **(B)** Sagittal view. **(C)** Transverse view.

This bending stress decreases as the focus of attention moves from the hip to the knee where it is replaced with more nearly compressive stress (Fig. 5-10A). As activities such as gait or stair climbing are added to the calculation, more "out of plane" loads are present, which tend to bend and twist the femur (Fig. 5-10B and Fig. 5-10C).

3 STRESS DISTRIBUTION AFTER TOTAL HIP REPLACEMENT

Implantation of a total hip replacement alters the pattern of stress distribution in the femur and pelvis. First, shear forces that were considered negligible in the normal joint are now significant (as much as 100 times greater than cartilage against cartilage) and can produce torques that can loosen a prosthesis where it attaches to bone. Second, the size and position of the contact area between the prosthesis and bone, which varies considerably between different implants, is crucial in determining the type and magnitude of the stresses developed. Finally, the materials from which the

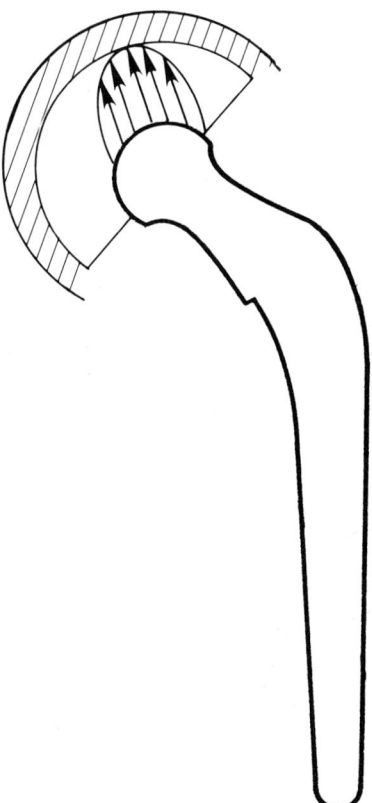

Fig. 5-11. The general stress distribution in the acetabular component due to joint force is mainly compressive.

replacements are made and the size and shape of the components will determine what stresses are placed on the bone.

In ideal circumstances, prosthetic components are in total contact with the supporting bones. Normal forces developed at the joint surface produce a similar pattern of compressive forces radiating into the prosthetic acetabulum and concentrating in the femoral head. If a classic all-polyethylene acetabular component is used with cement and if the classic technique of removing the subchondral bone of the acetabulum is followed, the acetabular prosthetic material will be stiffer (ie, less compliant) than the supporting bone (Fig. 5-11). Greater than normal compressive stresses will arise in the superior area of the cup while less than normal compressive stresses will arise in its medial aspect. In current joint reconstruction practice, changes in technique to preserve the subchondral bone and to support components without cement make generalizations about modern prostheses impossible, except that recognition of the above-noted changes in stress distribution lead to design changes from the classic technique.

When the femoral component contacts the acetabular component centrally, there is a minimal amount of bending of the acetabular component. As the resultant joint force acts more toward the periphery of the socket component, as it does in activities such as climbing stairs, there is a greater tendency to greater deformation (Fig. 5-12). This deformation can lead to failure of the polyethylene either by catastrophic breakage or wear. If the femoral component ball is smaller than the corresponding acetabular component socket, the resultant force will remain the same, but contact area will decrease, resulting in increased contact stress in the polyethylene and a tendency to central apical bending (Fig. 5-13). If the femoral ball is slightly too large for the acetabular socket, local stresses at the contact surface are again increased, but apical bending is eliminated or may even be reversed (Fig. 5-14). Metal support for the plastic component can decrease these de-

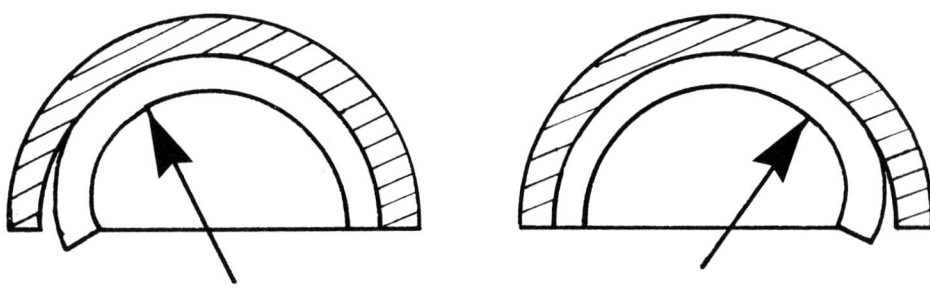

Fig. 5-12. Under an oscillating joint force the acetabular component has a tendency to be bent (exaggerated in this figure).

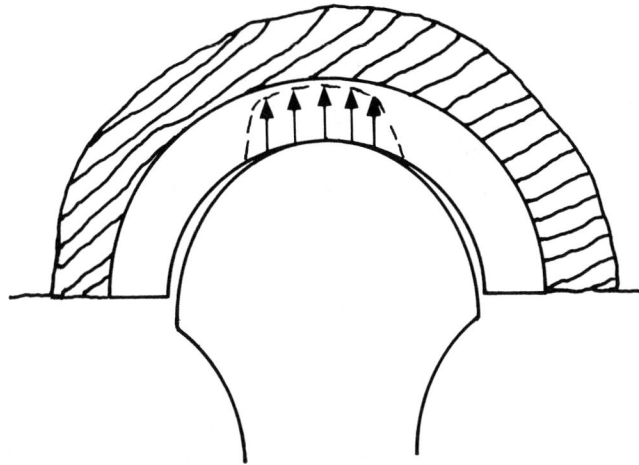

Fig. 5-13. Stresses increase when the femoral head is smaller than the acetabular component.

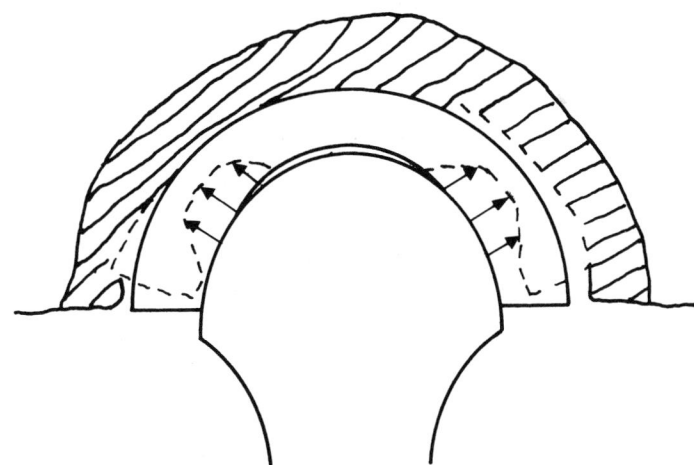

Fig. 5-14. When the femoral head size exceeds that of the cup, the stresses are concentrated at the periphery of the cup.

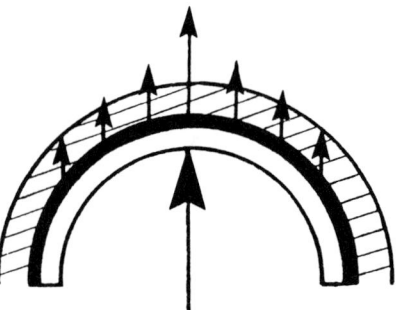

Fig. 5-15. A metal back constraining the polyethylene increases the stresses.

formations by firmly bonding the stiffer metal to the outside of the socket. The increased stresses then appear in the metal, which is more capable of withstanding them (Fig. 5-15). Increased stress may, however, then be transferred to the underlying bone cement, which may not be able to withstand cyclic loading of increased magnitude.

Whether one considers local contact stresses at the surface or overall stresses within the cup, the effect of prosthetic materials cannot be considered as static and unchanging. Since joint forces are intermittent in magnitude, and the resulting vector continually shifts in position with activity, the magnitude and pattern of the stresses and strains developed in the material are also continually changing. Thus an additional fact to be considered in the design of a prosthesis is the fatigue life of the implanted material. To add one more degree of complexity, the supporting bone cannot be considered as having static properties. Bone is a living structure that continually removes and adds elements in response to applied loads. This remodeling activity can, over time, either increase or decrease stresses and strains in implanted components.

4 PROSTHETIC ACETABULAR STRESS DISTRIBUTION AND JOINT SHEAR FORCE

In addition to compressive joint forces, the acetabulum is subject to and transmits shear forces created at the joint surface. Unlike the normal cartilage surfaced articulation, which has almost no frictional resistance, the prosthetic low-friction prosthesis bearing has a coefficient of friction that can be 40 to 50 times greater than that of the normal joint. This frictional resistance creates modest shear forces at the interface. Although these shear forces are much lower than the existing compressive forces, they must be included in any consideration of the mechanical effects of joint replacements. The role of shear forces on the acetabular component affects the contact

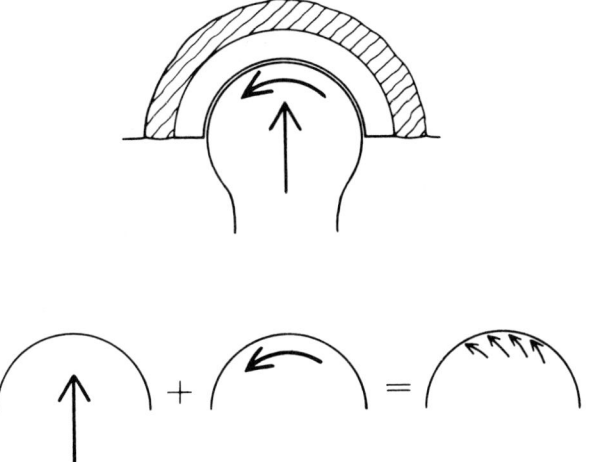

Fig. 5-16. The frictional (shear) forces are, at the surface, perpendicular to the compressive forces. The shear forces tend to alter the angle of the resultant stresses toward the direction in which the joint is moving.

stresses of the material in the vicinity of the joint and the stress distribution pattern throughout the material.

Shear forces alter the angle of stresses generated toward the direction of movement because near the surface of the joint shear forces are perpendicular to compressive forces (Fig. 5-16). The material being compressed in one direction must deform; tensile stresses and strains result (Fig. 5-17). This effect, in theory, decreases the further away from the surface one examines the stresses. If the shear forces are of sufficient magnitude, the material being compressed in the direction of movement could deform in that direction.

Polyethylene in retrieved components can be noted to have deformed. This deformation, which is not wearing away of material and occurs slowly over time, is called cold flow or creep. Such a phenomenon happens only to materials that are sufficiently plastic or deformable (Fig. 5-18). The material thins out where the compressive forces are maximal and accumulates where the compressive stresses are minimal. If a less deformable material is used

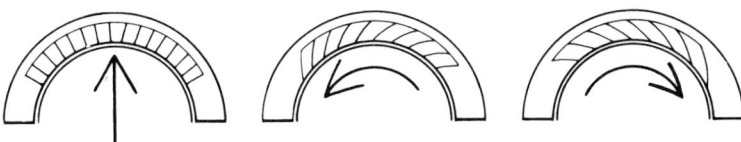

Fig. 5-17. Materials being compressed in one direction must deform. Tensile stresses and strains result. The reversal of motion sets the stage for creep or tensile fatigue of the acetabular component.

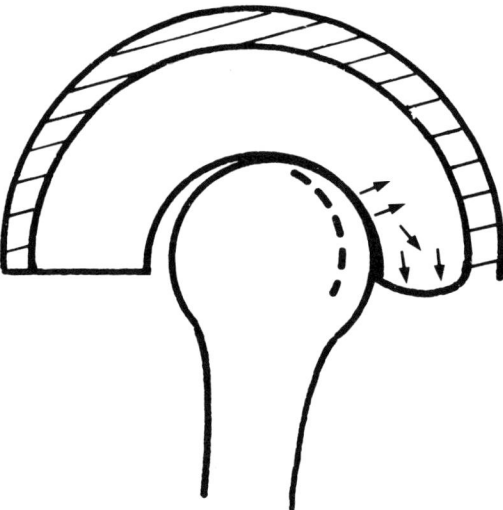

Fig. 5-18. Creep, or cold flow, in polyethylene.

or if forces are applied at a rapid rate, tensile strains break the material apart rather than causing the creep deformation.

5 STRESSES IN THE ACETABULAR CEMENT-BONE INTERFACE

The consequences of the stress distribution pattern in the acetabular component of a joint replacement are threefold:

- Possible mechanical failure of the component. Catastrophic failure of acetabular components has not been a problem except in association with loosening of the component and loss of support of the underlying bone.
- More energy transmitted. Prosthetic replacements do not absorb energy as well as a normal joint because more energy in the form of higher stresses and higher rates of loading is transmitted to the peripheral bony bed.
- Alteration in forces transmitted. Perhaps the most important consequence is the alteration in the forces transmitted to the interface between prosthetic components and host bone.

Stresses generated at the cement-bone interface surrounding the acetabular component are obviously the result of the compressive and shear forces at the prosthetic joint surface altered by their transmission through the acetabular component. (Recall that no tensile forces can be generated at the prosthetic joint articulation.) In general, because joint forces are directed predominantly superior and medial, this area of cement and bone is subject to compressive stresses. However, since the medial and lateral aspects of this interface are more parallel to the direction of major (compressive) force transmission, these areas are subject to shear forces (Fig. 5-19). When load is markedly eccentric in the acetabular component, because of the bending or rocking of the socket, tension forces can be present at the interface.

The greater the concentration of forces in the superior direction, the greater the shear forces on the medial and lateral sides. Such shear forces can be almost as high in magnitude as the resultant joint force itself. The greater the eccentricity of the load on the acetabular component, the higher the tensile loads on the opposing interface. Because compressive forces are intermittent, the shear and tensile forces they create oscillate as well. As long as the prosthetic surfaces retain their usual frictional resistance, they contribute only slightly to the cement-bone interface stresses by slightly increasing overall shear.

In general, the direction and magnitude of forces on bone adjacent to cement are similar to those in cement. Two factors, however, alter stresses in bone. First is the relationship between type of force and area of contact between prosthetic material (prosthesis and/or cement) and bone. Compressive force transmission requires only contact between cement and/or component and bone. However, in areas where there is no interdigitation of the bone with the prosthesis, either through cement or an "ingrowth" surface, shear and tensile forces developed are not transmitted to bone. Thus

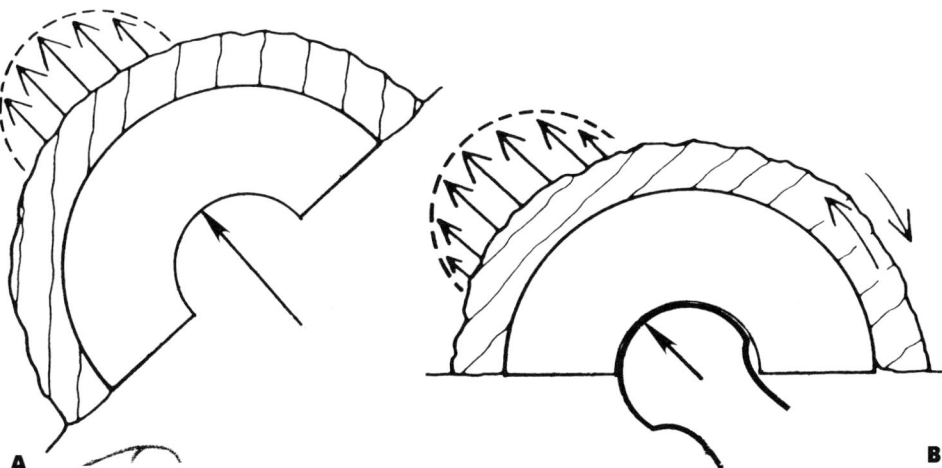

Fig. 5-19. (A) The superior and medial aspect of the bone-cement interface is primarily subject to compressive stress. **(B)** The medial and lateral aspects of the bone-cement interface are subject to shear.

Fig. 5-20. Adding pegs, or screws, to the acetabular fixation increases its resistance to shear.

the area available for transmission of compressive force can be considerably greater than that available for shear tension transmission.

The second factor that enters into development of stresses created at the bone-prosthesis interface is the architecture of the bone itself. Cancellous bone has a porous structure with considerable variation from one area and patient to another. In areas where bone is very dense, each trabeculum will undergo relatively less stress than in areas where bone is very porous. Furthermore, the trabeculae may be in an inappropriate orientation to carry the load of the prosthetic device, and vigorous remodeling may be necessary to reorient the bone. During such a reorientation, bone is initially lost and the trabeculae that remain must carry additional stress.

Mechanical failure on the acetabular side generally occurs at the interface between prosthesis and underlying bone. The surgeon and designer may limit the magnitude of the stress by (1) ensuring that contact between implanted materials and bone covers the largest possible area, (2) providing an immediate interlock between prosthesis and bone (by interdigitating cement, screws, pegs, and so forth with cancellous bone) to take the shear and tensile loads (Fig. 5-20), and (3) recognizing that the interface between the prosthesis and bone will initially weaken as the first stage of remodeling removes bone.

6 STRESS DISTRIBUTION IN THE FEMORAL PROSTHESIS

Compressive forces developed at the femoral prosthetic joint surface produce compressive forces in the prosthetic femoral head as in the normal joint. The magnitude and distribution of the stress may be altered by the factors cited in their discussion of the acetabular side: contacting area, location of contact, geometry, and type of material employed. In the neck and shaft of the femoral prosthesis, the compressive force eccentrically located at the base of the spherical head creates a pattern of predominately compressive

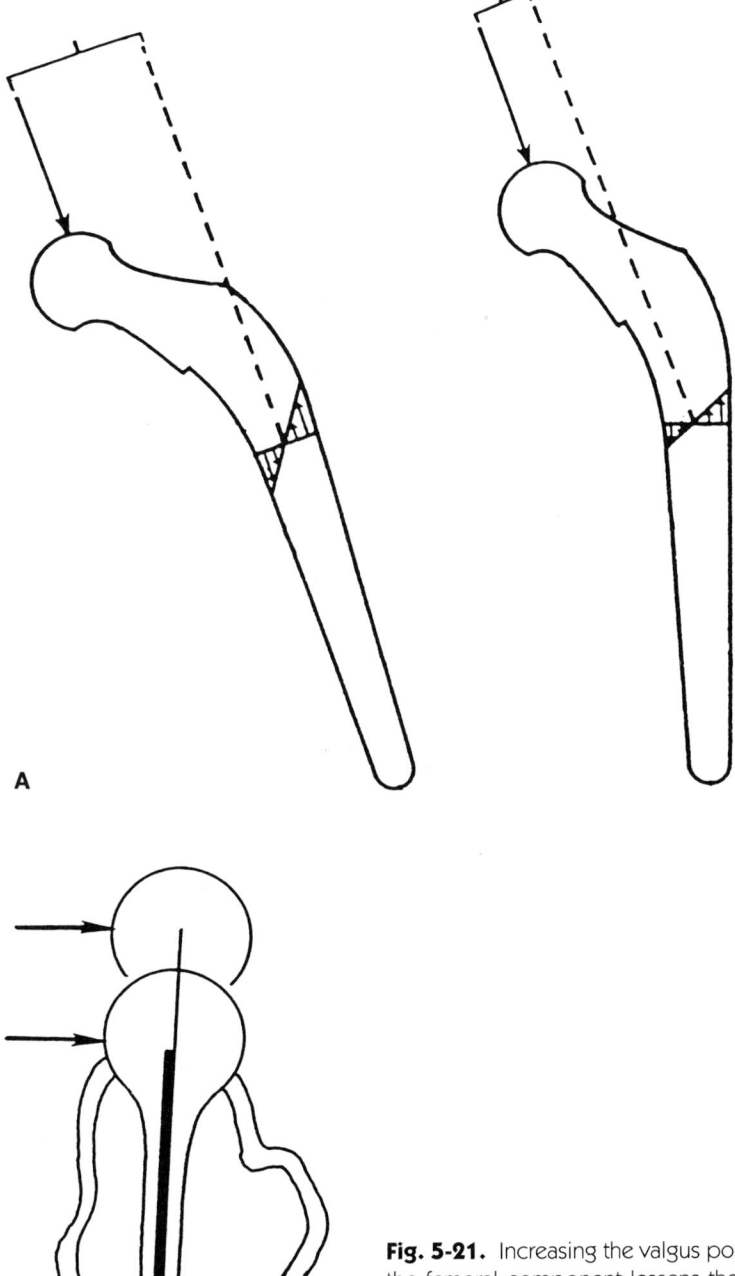

Fig. 5-21. Increasing the valgus portion of the femoral component lessens the lateral offset of the head from the shaft. **(A)** The more valgus stem shows lower stresses and in **(B)** the torque at the fixation interface is less with the more valgus position.

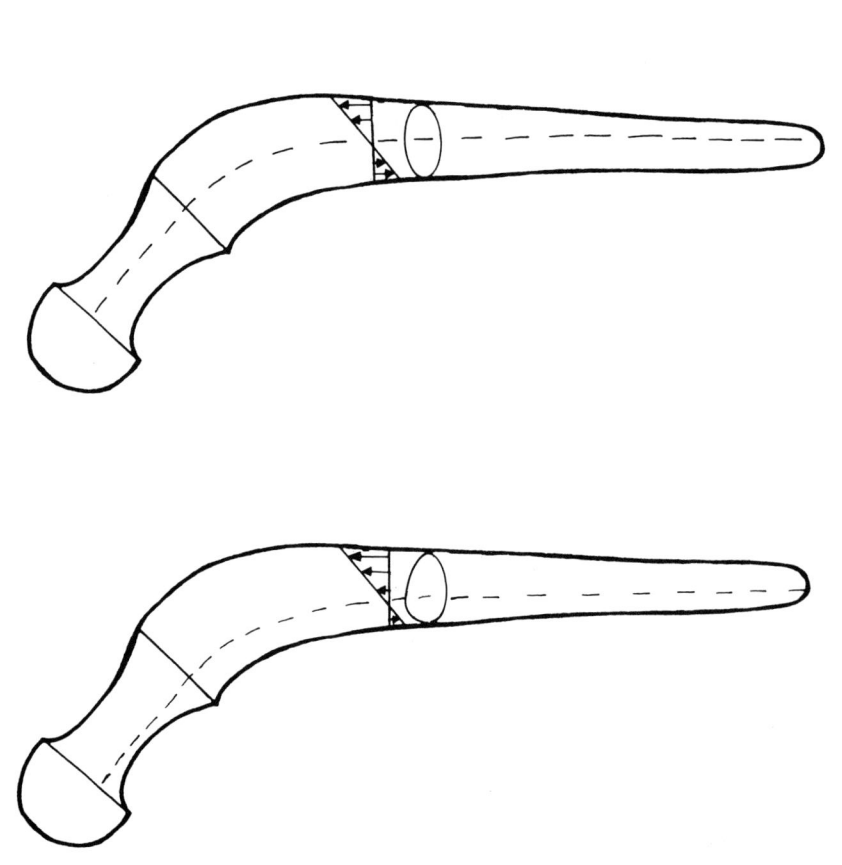

Fig. 5-22. The amount and distribution of the cross-sectional areas of the femoral component shank play a significant role in stress distribution in the stem.

and bending stresses similar to those noted in the normal femoral neck and trochanteric regions. Resultant force produces maximum tension on the anterior portion of the neck and maximum compression on the inferior portion because the load is directed from superior to inferior and medial to lateral as it is in the normal. While the resultant force in the transverse plane alternates from anterior to posterior and the reverse, the femoral head probably experiences a greater force directed posteriorly because the flexed position of the hip as it takes load minimizes anterior forces and increases posterior forces. Patients with hip prostheses frequently do not gain full extension of their hips after surgery. In those who do not, the posteriorly directed forces are more important than the anterior forces. This resultant force carries distally to produce bending and torsional stresses in the stem of the prosthesis.

Changes in geometry of the femoral prosthesis to alter the neck length (ie, the eccentricity of the compressive loading forces in relation to the axis of the neck) modify the bending and torsional stresses in the stem. Reducing neck length or designing a prosthesis with a more valgus orientation effectively decreases the offset of the prosthetic head relative to the stem and creates a more uniform compressive stress pattern because it decreases bending and torsional stress (Fig. 5-21). This effectively decreases compressive stress on the medial side and tensile stress on the lateral side of the stem. The magnitude of the stresses within the stem also depends on the cross-sectional area of the stem and on the distribution of the material in reference to its neutral axis (ie, the moment of inertia of the stem). The greater the amount of cross-sectional material, the lower the stress per unit volume as the total force is distributed over a greater area. The greater the cross-sectional area medial to the neutral axis, the lower the compressive stress and the higher the tensile stress (Fig. 5-22).

The shape of the stem of a femoral component (or any implant, for that matter) can serve to increase local stress by creating a stress concentrator. As was found relatively early in the history of hip replacements, a sharply angulated stem can concentrate stress sufficiently to lead to fatigue failure under the intermittent loading of normal patient activity. Ultimately, the strains that develop in the prosthesis as determined by the material properties of metal (modulus of elasticity, ultimate tensile strength, and fatigue strength) and the geometry of the component determine whether a given component will survive in vivo.

7 STRESSES IN THE FEMORAL PROSTHESIS SECONDARY TO FIXATION

The femoral component is subjected to stresses from compression, bending, and torsion. Bending and torsional stresses are the most potentially disastrous. Bending combined with torsion tends to concentrate tensile forces laterally. Because loading of the hip is intermittent, these forces (which are less than those required to cause immediate failure of the prosthesis) can lead to initiation and propagation of cracks, which over time can lead to stem fracture. The manner and degree to which the femoral component is constrained (ie, fixed) drastically alters the pattern and magnitude of the stresses developed in and about its neck and shaft. If a prosthesis is supported distally only, the bending and torsional stresses are concentrated on a portion of the stem with a smaller cross section and is thus less able to withstand stress. If a stem is supported more proximally, a larger cross section is available to carry the load. Cemented stems that have loosened proximally have broken in vivo. Failure can occur either in the stem or in the material used to fix the prosthesis to the bone (either with or without cement) (Fig. 5-23).

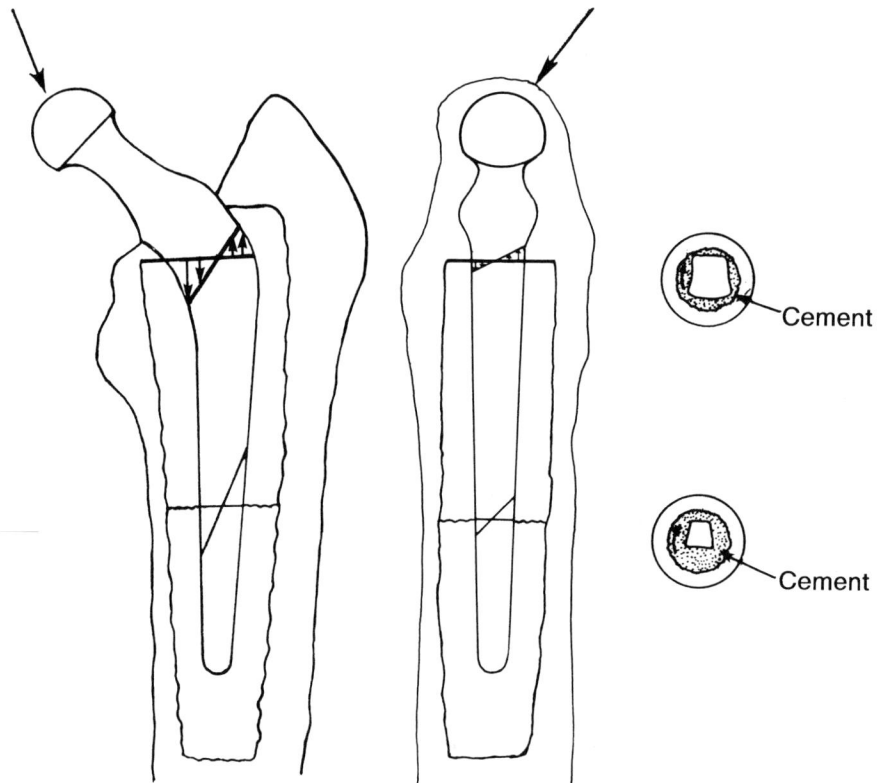

Fig. 5-23. A prosthesis with proximal support will have considerably different stresses from one supported distally.

8 FORCES AND STRESSES IN THE CEMENT AND BONE SURROUNDING THE FEMORAL PROSTHESIS

Stresses created in the area of fixation on the femoral side are highly dependent on the shape, size, and position of the prosthesis, as well as on the location, orientation, and amount of contact between it and the surrounding fixation to bone. Though joint shear forces contribute somewhat, the compressive forces at the joint contribute most to the bending and torsional moments created.

The advantage of a low friction over a higher friction design is that less muscular force is required to move the opposing surfaces relative to one another. This lessens the force on the joint in both shear and compression. Regardless of how the bending and torsional moments are created, if loosening of the prosthesis is to be prevented, the forces implied by the moments must be resisted by forces generated mainly by the fixation interface proximal-medially and distal-laterally.

Attempts to reduce the bending moment by decreasing the degree of varus in the prosthesis or decreasing the neck length chosen (ie, decreasing the lateral offset of the femur relative to the center of rotation of the hip joint) increase the compressive component of the forces transmitted down the prosthetic stem. Consequently, this increases compressive stresses generated in the fixation region and decreases stresses generated by bending and torsional forces. While overall stress is decreased by this design, effectiveness of musculature about the hip depends on lateral offset, which is decreased by a more valgus design. Prosthetic design or insertion techniques that lead to too little offset lead to poor function.

Improving the ability of the fixation region to carry the stresses generated can allow an appropriate offset. The mechanical load transfer of cement to the bone and femoral shaft (by rendering the cement less viscous, plugging the canal to provide a closed space, and injecting the cement under significant pressure) can be improved. Techniques to improve the mechanical properties of cement itself by vacuum mixing and centrifuging improve load transfer in the cemented prosthesis. Cementless prostheses depend on an intimate fit of the femoral stem in the canal to resist the loads applied. A pressure fit technique in which a slightly undersized hole is created in the bone into which the prosthesis is forcibly inserted is commonly used for cementless implants. The resulting interference fit is intended to provide load transfer from prosthesis to bone and to prevent migration of the stem.

Once a prosthesis has been implanted, bone adaptation to the new mechanical environment must occur. This is likely the most important determinant to long-term success. Too stiff a prosthesis can lead to the stresses being concentrated at the end of the implant with resultant disuse osteoporosis in between. Too elastic an implant can lead to stress concentrations that the bone cannot tolerate and from which the bone resorbs or fractures. The search for the right combination of stiffness and fixation has resulted in many different designs, all of which succeed in many patients but fail in others. While we can calculate the forces of compression, tension, torsion, and shear, the reaction of an individual bone to these forces is not as of yet predictable and represents the greatest challenge of biomechanics.

9 KNEE JOINT STABILITY

Joint stability is dependent on forces acting in a given direction. The forces cause motion (ie, joint range of motion) or are prevented from causing motion (ie, joint stability). When motion is caused, its speed and total amount must be controlled. Three mechanisms at each joint maintain stability: cartilage and bony geometry, muscular action, and ligamentous restraint. The degree to which each contributes varies from joint to joint.

The ball and socket geometry of the hip prevents any translational motion but allows rotation around a central fixed axis. At the extremes of motion, the capsule and constraint of the femoral neck abutting the acetabular margin limit motion. Muscle contraction provides dynamic stability for the hip except in extension, in which position the strong anterior capsule limits hyperextension. Implantation of an appropriately positioned total hip replacement does little to alter normal muscular control. Formation of a tough fibrous pseudocapsule replaces the anterior capsule. Unless there is malposition of one of the components that leads to impingement of the neck of the femoral prosthesis against the acetabular component or bony acetabulum, stability of the replaced hip is ensured.

In the knee joint, the factors that determine stability are somewhat different from those at the hip. The knee is a flexural linkage—its major motion is to extend to take the load of standing and walking and to flex to allow gait and other activities. Flexion-extension, abduction-adduction, and internal-external rotation occur simultaneously around three axes of rotation that are not fixed (Fig. 5-24). Motion can be translational as well as rotational. Some motions are mandatory during activity, while others are voluntary under neuromuscular control.

At the knee joint, various biomechanical adaptors allow some motion in certain directions but limit their range. Specific ligaments connect the femur and tibia. Muscles control the speed of such motions, especially when high torques are present (such as when bending, jumping, or going up and down stairs). Changing lever arms maximizes the counteracting torques muscles can develop. Thus stability is achieved by a combination of surface contours, ligaments, and joint capsules that both permit and limit motion while providing optimal lever arms for the activated muscles.

The relative importance of each factor varies depending on the motion

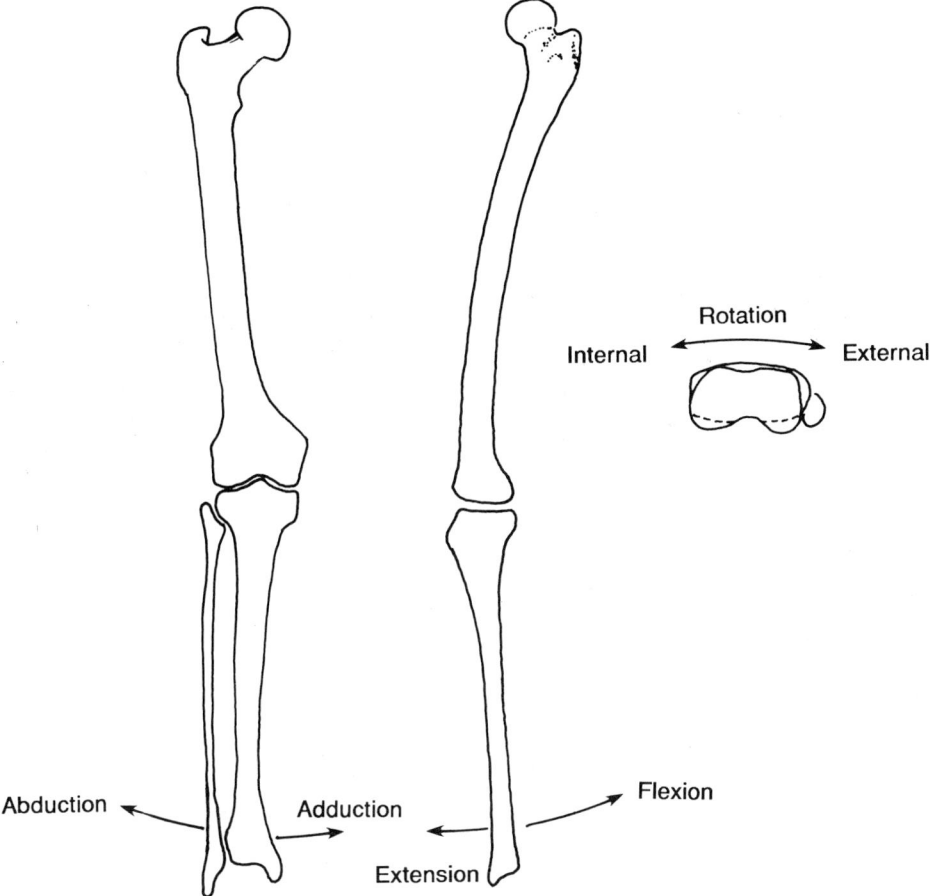

Fig. 5-24. The definitions of movement in abduction, adduction, flexion, extension, and rotation.

studied. For example, in the frontal plane, medial-lateral translational motion is restricted by the conformity between the femur and tibia in the region of the tibial spines (Fig. 5-25). Varus force developed by action of body mass through the contact point with the floor is limited by tension developed passively in the lateral collateral ligaments and actively by muscular contraction that tightens the iliotibial band (referred to as the lateral muscular stay) (Fig. 5-26). In the sagittal plane, high torques are created by body mass tending to flex the hip and knee. The demands of controlling these large torques and producing counteracting ones require large muscle masses acting through continually changing lever arms provided by the nonspherical (ie, multi-axis) femoral condyles (Fig. 5-27).

To provide for the changes in lever arms, the contact area of the femur with the tibia changes, moving generally backward with more flexion and

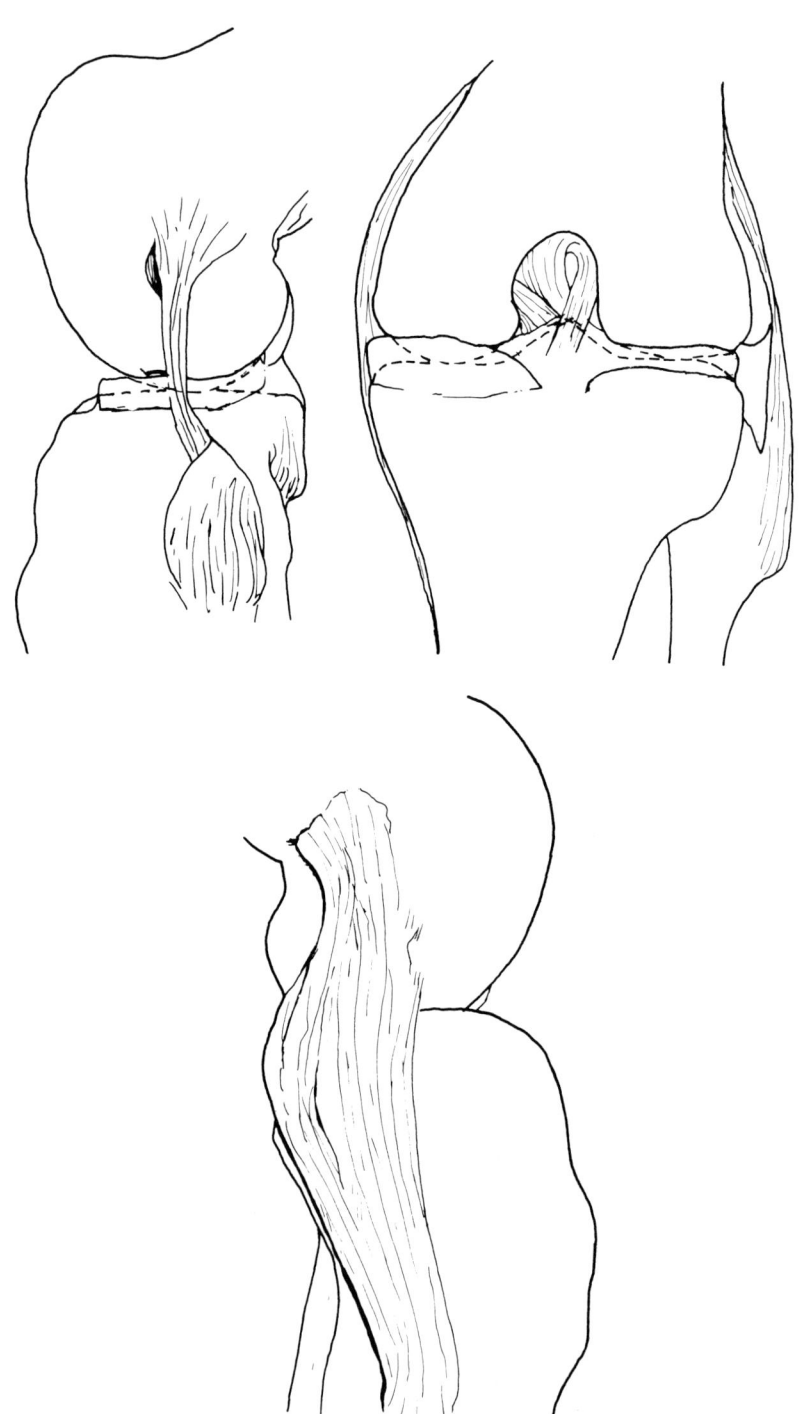

Fig. 5-25. Medial-lateral translation of the knee is limited by the tibial spines.

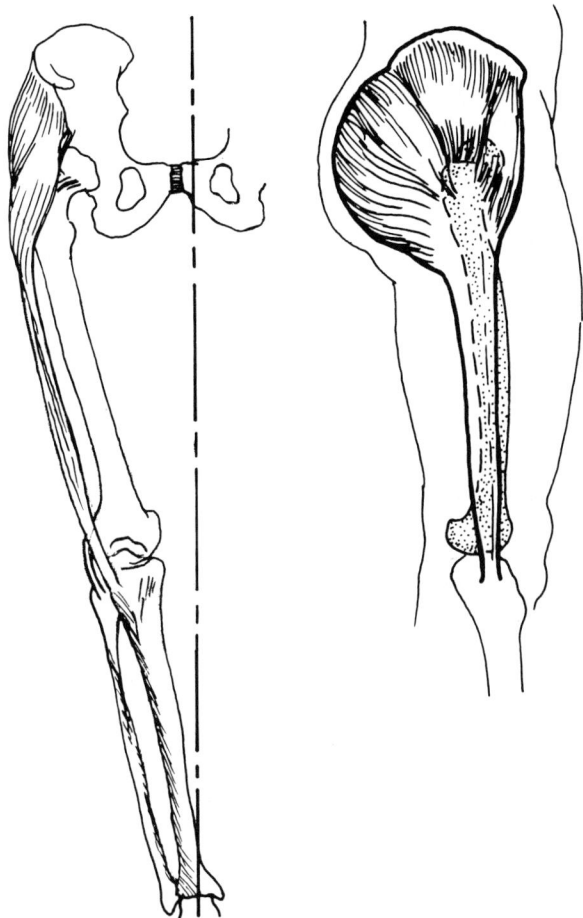

Fig. 5-26. Varus force is limited by passive tension developed in the lateral collateral ligaments and by the active contraction of the lateral muscular stay.

forward with more extension. To allow the moving contact area, conformity between tibia and femur is limited and anteroposterior translational motions are limited in the extreme by the strong cruciate ligaments, the collateral ligaments, and, to a lesser degree, the posterior capsule. Menisci attach to the tibia and slide anteriorly and posteriorly with the contact area to add more stability.

Change in rotational position of the tibia in the transverse plane makes the contact area between the femur and tibia relatively more anterior or posterior in one compartment than in the other. Extremes of this motion are limited by ligaments and by the conformity of the surfaces. Instantaneous control is probably a neuromuscular function with differential con-

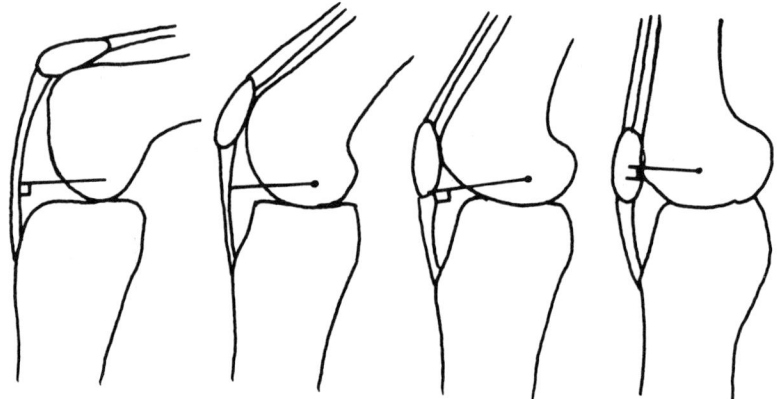

Fig. 5-27. The lever arms of the muscles about the knee change with knee position. Note the decrease in lever arm of the quadriceps with extension, as the torque required from that muscle is decreased.

traction of hamstring and popliteus muscles positioning the tibia and resisting rotational torques.

Deterioration of the knee joint has a dual effect, it causes a change in the bony contours of the joint, and it alters ligament lengths, usually shortening them with scar. Surgical replacement of bony contours of the knee is more difficult than replacing the relatively more simple hip joint. Design of the knee prosthesis must take into account the changing instantaneous axis of rotation of the joint. Instrumentation must place the prosthesis in such a position that ligamentous structures and the prosthetic contours work in harmony, providing motion and changing lever arms for muscle function.

The spectrum of sizes of human beings and thus of their knee joints makes some design compromises necessary. Add to this the fact that the forces normally acting through the knee joint are probably beyond acceptable limits for prosthetic materials now available. One can see the difficulty in coming to a final design specification for a total knee system. Several hundred designs for knee replacement have been fabricated, but, while some radically different designs persist, most have taken on a "condylar" look with varying degrees of conformity between the tibial and femoral component (Fig. 5-28).

10 KNEE JOINT STRESSES

In a patient who has only minimal surface damage to the knee joint, ligaments should be intact and only minimally damaged. In this circumstance, merely replacing the surfaces with two unconnected parts with the same conformity as the normal knee should suffice for replacement. Under such a circumstance the degree of stability actually achieved is dependent on (1) tightness of the ligaments when inserting the prosthetic components and

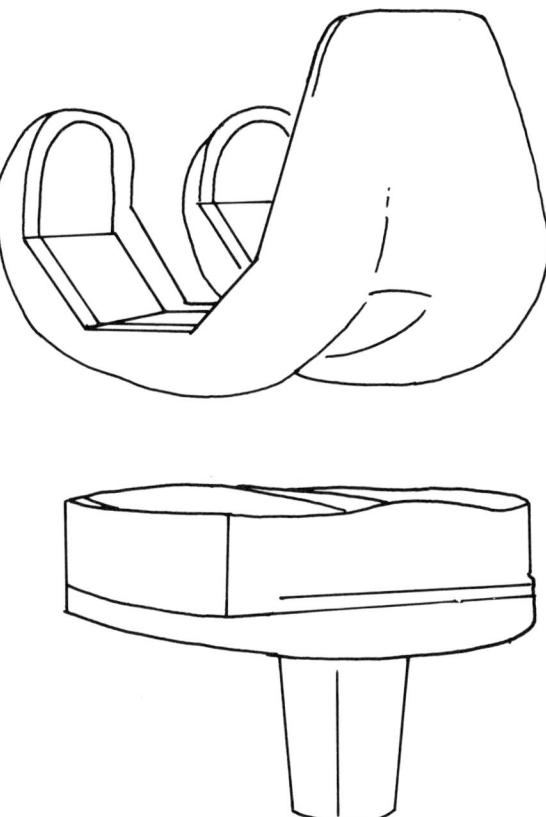

Fig. 5-28. Total condylar prosthesis design.

(2) surface contour of the components. While such a design is attractive in theory, the materials available do not make this approach feasible.

The materials commonly in use are, as in hip replacement, noncorrosive metal (cobalt-chromium or titanium alloy) and polyethylene. Polyethylene, even the ultrahigh molecular weight version, deforms (as any plastic would) under relatively low stresses. Continued application of stress causes continued deformation, called creep or cold flow. Therefore, the surface interface between the metal and plastic components of the prosthesis must be maximal to avoid exceeding the elastic limit of the polyethylene.

In the hip, unless there is deformation of the polyethylene component, maximal contact area occurs at all times within a normal range of motion. Since the knee normally changes the contact area between the femur and tibia throughout the range of motion (ie, it slides as well as glides), contact area must be limited in some parts of the range of motion. A design with femoral and tibial components of exactly the same radius provides for maximal contact area, but does not allow for the changing contact point. In such

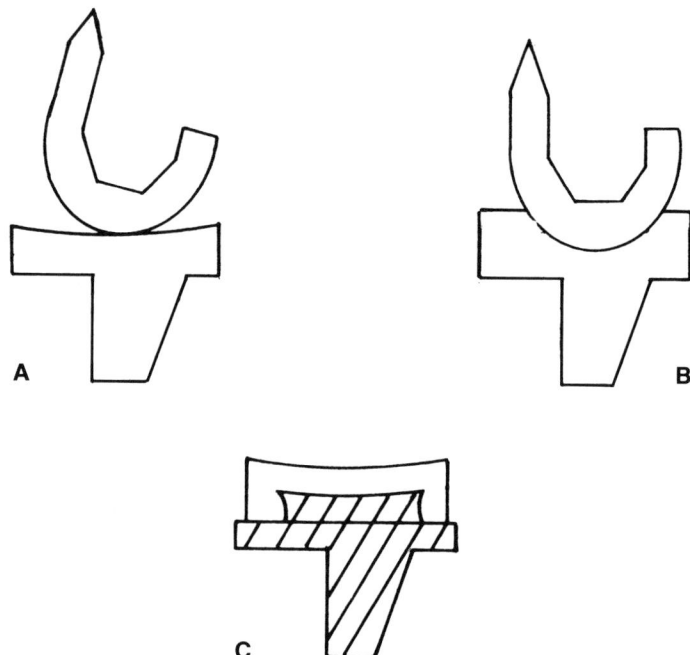

Fig. 5-29. (A) The conforming femoral-tibial couple with high contact stress and maximal freedom of glide and **(B)** the conforming femoral-tibial couple showing the opposite. **(C)** Metal support for polyethylene may allow it to take higher stresses.

a case, all translational motion (whether caused by axial rotation or by change in contact point dictated by ligaments) would be restrained by the component's walls (Fig. 5-29).

Forces developed in the walls would be significant and the area of component to carry the loads small. The result can be cold flow of polyethylene or fracture of the component from either a single insult force beyond the ultimate strength of the component or repeated less-than-ultimate yield forces that can lead to fatigue failure. Making the tibial components flat, like a sled, allows more freedom of motion and allows the knee to slide. However, such a design leads to high stresses at the contact point between femur and tibia. Such contact stress can lead to cold flow or fracture as described above. In response to this problem, many tibial polyethylene components are now supplied with a metal back and frequently a supporting skeleton for the plastic.

Demand on designers of knee prostheses with regard to contour of surfaces is more critical than for hip prostheses. Balance between the competing interests of surface geometry, ligamentous balance, material limits, adequate sizes within an inventory that hospitals can afford, and cost of the implant in a world of shrinking resources make knee replacement design a difficult proposition. It is not surprising that most designs are beginning to look monotonously similar.

Designs, produced with the supposition that the posterior cruciate ligament should be retained, tend to have less conformity between femur and tibia relying on the retained ligament to prevent posterior subluxation of the tibia on the femur (Fig. 5-30). Designs that assume posterior cruciate sacrifice most often show conformity by means of a projection from the tibial component. The projection contacts an area on the femoral component to prevent posterior subluxation of the tibia and to push the tibia forward (in effect rolling back the femur on the tibia) as the knee flexes. Most designs assume that the anterior cruciate ligament will be nonfunctional because of the disease process and count on its absence.

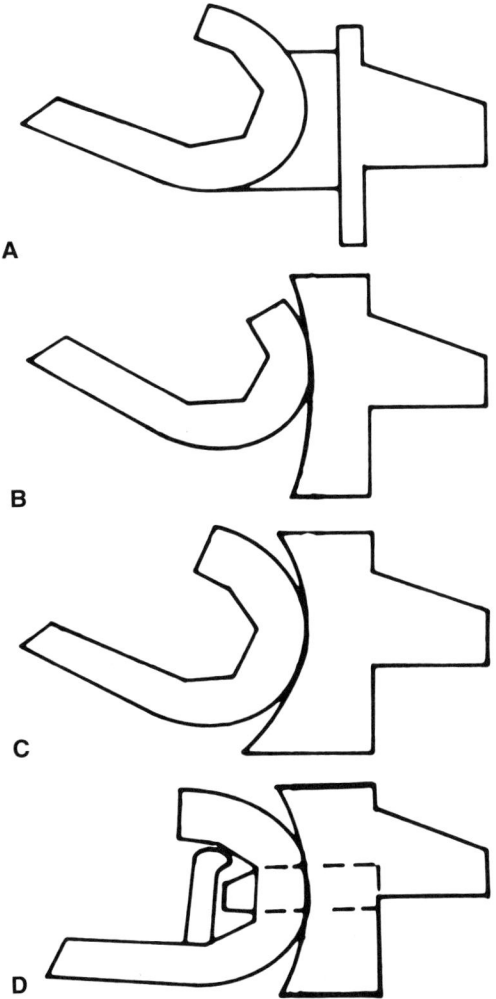

Fig. 5-30. (A) Posterior cruciate substitution with a central spike projecting from the tibial component. **(B)** Posterior cruciate substitution with a large anterior lip on the tibial component. **(C)** Posterior cruciate sparing design with less conformity. **(D)** Meniscal bearing design allows separate motion between femoral and tibial components against the polyethylene bearing.

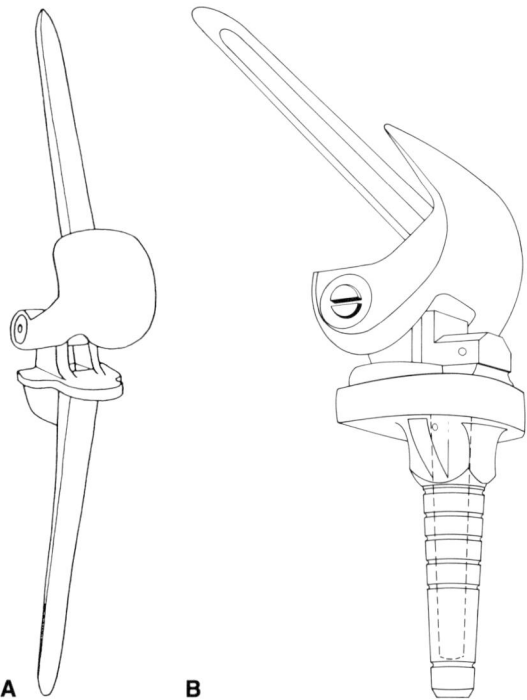

Fig. 5-31. (A) Single-axis hinge prosthesis allows only flexion-extension. **(B)** Multi-axis hinge prosthesis allows flexion-extension, internal-external rotation, and some abduction-adduction.

Several designs have moveable polyethylene meniscal bearings that have complete conformity between the femur and the polyethylene and a large surface area over which the polyethylene piece can glide on a polished metal tibial component. These types of components have the advantage of a large area of contact while allowing independent translation of the bearings to answer the theoretical need for lack of constraint and conformity.

Total knee joint replacement has been performed for at least 25 years. Some designs have been abandoned because of unsatisfactory results, and the differences between available designs today are vanishingly small. There appears to be no functional advantage to cruciate preservation except for gait analysis evidence of more normal motion during stair climbing. Meniscal-bearing polyethylene components have not shown a functional advantage. However, some lessons in practical biomechanics can be learned from the devices that have been associated with failure. It used to be considered convenient to divide total knee designs into constrained and nonconstrained. Such a division became blurred as most knee joint prostheses became semiconstrained. A maximally constrained prosthesis would be a uniaxial hinge placed in a knee joint to mimic only the flexion-extension motion of the joint (Fig. 5-31). Such joints had a higher than acceptable amount of loosening of the prosthesis at the bone-cement interface and a much higher than acceptable number of fractures of the metallic stems of the implants. These prostheses failed to provide for the complex motions of the joint, which led to fixation interface stresses beyond the limit of cement interdigitated with bone. Linked prostheses that allow functional flexion and extension and some abduction-adduction and axial rotation have improved results. Linked components with no intrinsic stability are still in use in patients because of bone loss associated with previous failed arthroplasty, infection, or tumor.

Many total knee failures occurred on the tibial side, with the tibial component sinking into the top of the tibia. This sinking generally occurred on the anteromedial part of the prosthesis. Computer modeling of the tibia with a component in place suggested that stress on bone beneath a pure polyethylene component was too high, and failure resulted from fracture of the bone and bone cement.

Metal backs, originally used to support the polyethylene and to minimize cold flow, were suggested to have the advantage of protecting bone from high stresses. Analysis with computer modeling, however, predicted that a metal back alone without the addition of an intramedullary stem would serve to concentrate the stress at the edge of the metal (Fig. 5-32). This proved to be true, and metal-backed tibial components without additional fixation have had a high failure rate and thus have been abandoned.

In the absence of long intramedullary stems with a fixed angle between stems, alignment of the limb following a total knee replacement is in the hands of the surgeon implanting a prosthesis. Alignment can markedly change the magnitude and type of forces as well as their distribution across the knee joint. Varus angulation (whether associated with lateral ligament

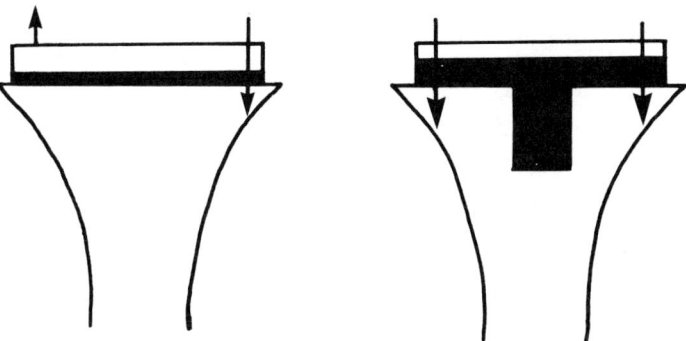

Fig. 5-32. The bolt in a hinge prosthesis, because of the varus strain on the knee in gait, is subjected to compressive stresses as shown. The stress laterally is less than the stress medially because of the sparing effect from the addition of bending to compression.

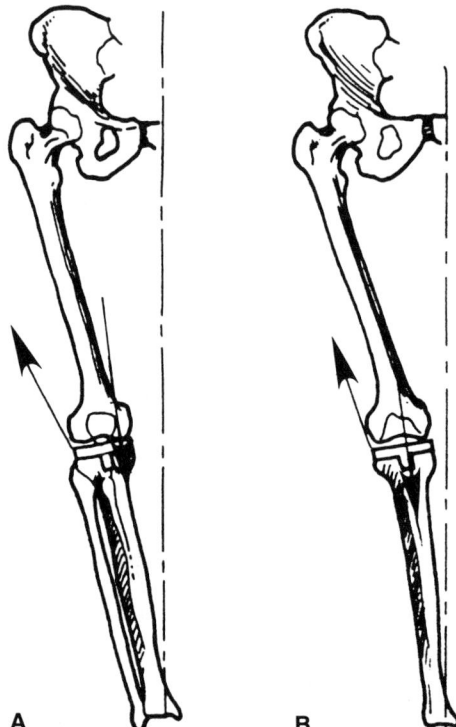

Fig. 5-33. Varus alignment of a total knee increases the compressive stresses on the medial side. **(A)** Varus alignment. **(B)** Physiologic alignment.

laxity or not) leads to increased compressive forces on the medial side (Fig. 5-33). Bending moments created by such an asymmetric load create increased compression on the prosthesis and bone on the medial side and an increase in tension at the fixation interface and ligaments on the lateral side. Valgus angulation has the opposite effect, causing an increase in compressive stresses on the lateral side and in tensile stresses on the medial side.

Some eccentricity in the application of load to the knee is unavoidable. The point of maximal compressive force varies with the activity being performed and with the rate of motion of the joint. High compressive loads on the medial, lateral, anterior, or posterior portions of the tibial component create tensile stresses on the opposite side of the component (Fig. 5-34). A means to neutralize these loads must be provided through cement, bone ingrowth, press fit, or screw fixation.

The patello-femoral component of total knee arthroplasty deserves some mention. High loads implied by contraction of the flexion-extension musculature greatly tax prosthetic materials. Loads on the polyethylene patellar component with limited contact are high enough to cause significant cold

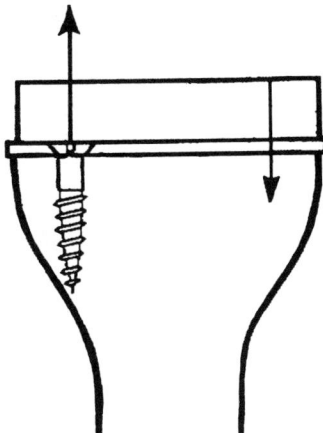

Fig. 5-34. Compressive stresses on one side of a tibial component can lead to high tensile forces on the opposite side. Neutralization of these tensile forces can be accomplished with a screw.

flow. Enlarging the surface area to prevent cold flow leads to a more constrained patello-femoral design and greater shear and tension at the fixation interface if the articulation must resist forces tending to dislocate the patella laterally (Fig. 5-35). Limiting these shear forces decreases contact area, but cold flow and wear can occur.

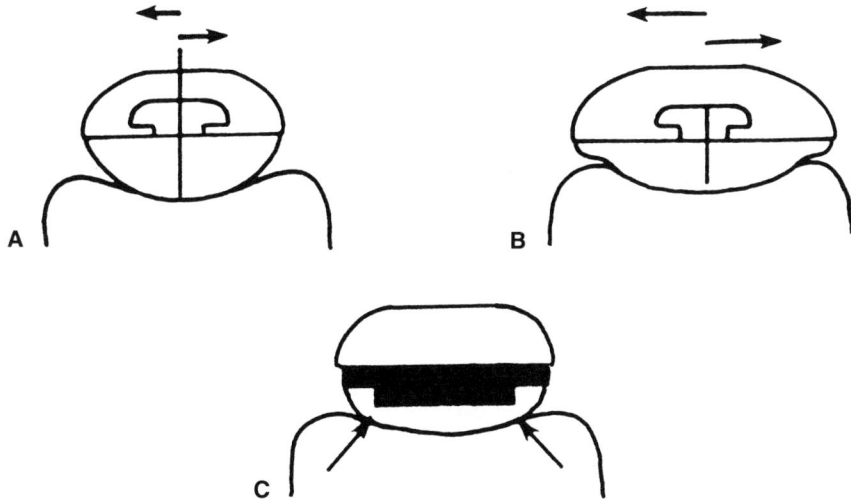

Fig. 5-35. High stress concentrations on the polyethylene tibial component can be associated with its significant deformation. **(A)** Minimal constraint minimizes shear at the fixation interface. **(B)** Increasing constraint increases the shear at the fixation interface. **(C)** Metal backs have led to failure when polyethylene was too thin between the metal back and the femoral component.

Attempts to place a metal back on the patellar component have led to dramatic failures. The limited size of the patella led to a decrease in thickness of the polyethylene to accommodate the metal back. The limited polyethylene thickness led to stress concentration by the metal beneath. Wear of the polyethylene exposed two metallic surfaces to abrasive wear in the patello-femoral articulation with significant consequences of loosening and pain. The compromise between area contact and lack of constraint is evident in most current designs, which accept some cold flow to prevent loosening.

11 WEAR

In our discussions of the motions and forces at the interfaces between prosthetic components, breakdown of the surface material—wear—has been mentioned several times. Discussion of this important subject has been left for last because it is relatively poorly understood; however, it is assuming an ever increasingly important role in the design of total joint prostheses.

Wear is basically the removal of material from the surface. An implant could, in theory, fail because of wear or because of changes in surface geometry caused by wear. Of considerably more importance, however, is the biologic reaction to wear debris. Debris derived from any source, if the particles are of the correct size, elicits a response from the host mesenchymal cells that leads to progressive resorption of bone at the interface and loosening of the prosthetic components.

There are four basic mechanisms of wear. The first, an abrasive type, occurs when an irregularity of one material, when moved over the surface of the other, scrapes out a path something like a plow through a dirt field (Fig. 5-36).

The second mechanism of wear is adhesive wear. This occurs when the irregularity of one surface comes into contact with that of the other and,

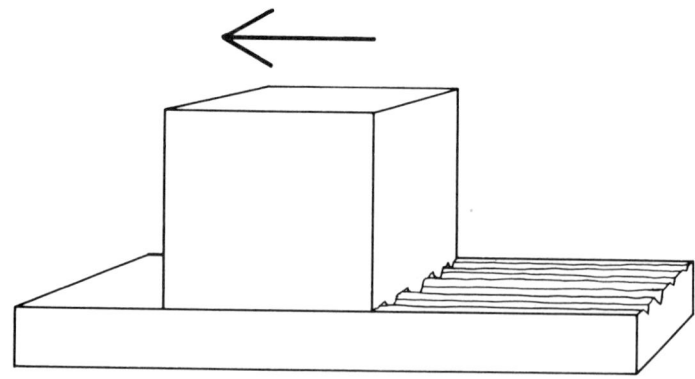

Fig. 5-36. Abrasive wear. One surface scratches the other.

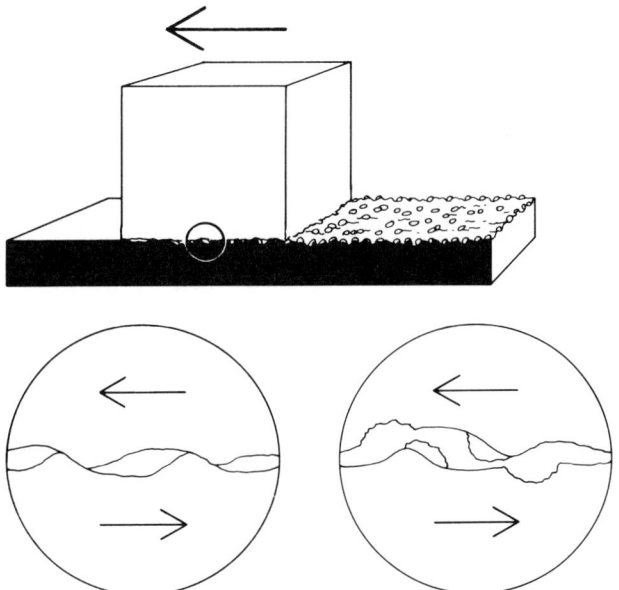

Fig. 5-37. Adhesive wear. The tops of asperities adhere and are broken off at their bases.

because of simultaneous compression, a chemical reaction causes one surface to adhere to the other. Continued motion breaks the two joint materials apart, but it does so at a different interface. The high spot on one surface is then prone to fracture. This may be likened to the ease with which stalactites can be broken off in caves or ice on the sides of houses in the winter (Fig. 5-37).

The third type of wear can also be considered an abrasive or adhesive type but of the third-body variety. Instead of one surface ploughing through or connecting to the opposite surface, an additional material introduced between the two surfaces operates in the same fashion. This can occur in total joint replacements from high density polyethylene (HDPE), cement, particles of porous coatings, or particles of ingrowth enhancement coatings such as hydroxyapatite. In joints that have their surface contours shaped so that concave surfaces are facing upward, gravity can draw debris into the joint rather than remove it. The differences between designs for the hip and for the knee reflect the higher tendency to retention of wear debris in the knee than in the hip (Fig. 5-38).

The fourth type of wear can result from nonuniformity of stresses at the surface, creating high compressive loads in certain areas and tensile loads at directly adjacent areas. The area of abrupt change from compressive to tensile stress is subject to high strain, which, if sufficiently large, can create fractures. Because of the constant motion of the joint, this stress concentration is changing over a certain small area, and, if many small fractures

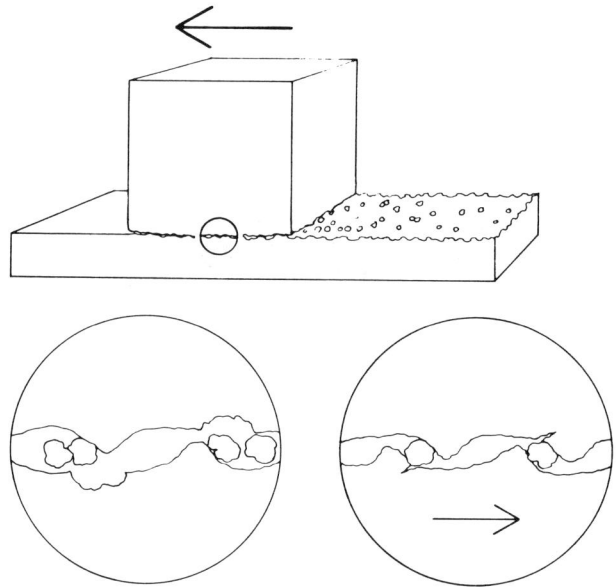

Fig. 5-38. Third-body wear: debris between the bearing surfaces acts as an abrasive.

connect, a piece of material can separate from its prosthetic source (Fig. 5-39).

All types of wear patterns in prostheses depend on stresses that are present in relation to the strength of the existing materials. In general, stresses in the knee are higher than are those at the hip because contact areas are smaller in the knee. Hardness of the material is a good parameter by which to judge the degree of wear. If two surfaces are of different strengths or hardness, wear occurs in the softer of the two. This can be altered if there

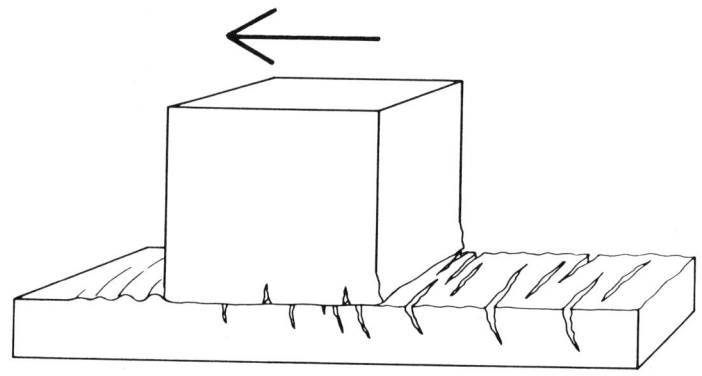

Fig. 5-39. Fatigue cracks due to repeated surface stress.

is a certain chemical affinity between the two surfaces because then wear may be accelerated by adhesive mechanism.

The harder the contact surface, the less the wear. Theoretically, less wear material is present with a metal-on-metal prosthesis than with a metal-on-HDPE prosthesis. If, however, the material is so soft that it deforms and the stresses created are absorbed by this deformation, no true wear occurs, only changes in surface geometry. Such is the case in cold flow of polyethylene.

Attempts have been made to decrease the wear of HDPE by rendering it harder with carbon fibers. The resulting material was indeed harder, but the reduced cold flow afforded by the carbon fibers led to higher contact stresses and the wear debris was not decreased. Bearings of ceramic materials against ceramic materials have a very hard bearing couple and have shown very low rates of wear; however, the difficulties in fabricating the pairs and in fixation of such a stiff material to the skeleton have limited their widespread use.

Appendix: Explanation of Units

There is general movement toward a unified international system of units. Most scientific journals have adopted this "International System of Units" (officially abbreviated "SI") and require that British units and mks units (meter-kilogram-second, which is a subsystem of the SI system) be converted to the official SI units for publication. The SI system is based on six basic units, five of which are relevant to bio-engineering type of activity. These units are length . . . meter (m); mass . . . kilogram (kg); time . . . seconds (s); temperature . . . Kelvin degree (°K); electric current . . . ampere (A). All other SI units are derived from these. Below is a list of the most important units (official SI units are italicized) and conversion factors for other systems.

Length: *1 meter (m)* = 100 centimeters (cm); 1 foot (ft) = 0.3048 m; 1 micron = $10 \times^2$ m; 1 angstrom = $0.1 \times 10 \times 9$ m; 1 inch (in) = $2.54 \times 10 \times^2$ m.

Linear velocity: *meter/second (m/s)*; 1 kilometer/hour = 0.2778 m/s: 1 mph = 0.447 m/s; 1 foot per minute (FPM) = $5.08 \times 10 \times^3$ m/s.

Mass: *kilogram (kg)* = 1,000 grams (g); 1 pound-mass (lbm) = 0.4536 kg; 1 slug = 14.594 kg; 1 ounce (oz) = 0.02835 kg.

Force: *newton (N)* = 100,000 dynes; 1 pound-force (lbf) = 4.4482 N; 1 kilogram-force (kg) [do not use kilogram for force, it is a unit of mass] = 9.8067 N; 1 gram-force (g) [do not use grams for force, it is a unit of mass] = $9.8067 \times 10 \times^3$ N.

Pressure: *newton per square meter (N/m^2)*; 1 pound per square inch (psi) = 6894.8 N/m^2; 1 normal atmosphere (760 torr) = 101,325 N/m^2; 1 bar = 10^5 N/m^2; 1 kilogram-force per cm^2 [do not use kg/cm^2] = 98,066.5 N/m^2; 1 mm of mercury = 133.32 N/m^2; 1 lbf/ft^2 = 47,880 N/m^2.

Energy, work: *joule (J)* = 1 newton-meter (N-m) = 10^7 erg; 1 foot-pound (ft-lbf) = 1.3558 J; 1 British thermal unit (BTU) = 1,054 J; 1 calorie (cal) [try not to use calorics] = 4.184 J.

Power: *1 watt (W)* = J/s; 1 foot-pound (ft-lb) = 1.3558 W; 1 horsepower (hp) = 745.7 W; 1 BTU/hr = 0.2931 W.

In this text we use the SI units, followed in parentheses by the British unit equivalent.

Glossary

Acceleration   Rate of increase of an object's velocity (meters or feet per second per second).

Anisotropic   Directionality of mechanical properties (ie, the material does not behave the same in all directions).

Annealing   Heat treatment used to render metals softer and more ductile.

Anode   In a battery or corrosion situation, the more reactive metal that dissolves (ionizes) and gives up electrons.

Area moment of inertia   The relative resistance to bending of a given cross section. The stress due to bending at any point is proportional to the bending moment and inversely proportional to the area moment of inertia.

Bending   Induction of curvature in the long axis of an object by the application of an eccentric force or bending moment.

Bending moment   (moment of a force) Measure of the bending intensity created by a force (obtained by multiplying a force by its lever arm).

Boundary lubrication   Separation of bearing surfaces by a film of lubricant that adheres to the surfaces themselves. Also dry friction where asperities of high points of the two bearing surfaces touch when rubbed. The latter definition is in a strict engineering sense and does not apply to a joint that always contains a film of fluid lubricant.

Brittle   Sustains little or no permanent deformation prior to fracture.

Buckling   Bending produced by vertical forces along the long axis of an object.

Casting   Fabrication of parts by melting and pouring into molds.

Cathode   In a battery or corrosion situation, the less reactive metal. It does not corrode.

Center of gravity   For analytical purposes, the point at which the mass of an object is thought to be concentrated. Point at which any object is balanced exactly.

Coefficient of friction A parameter used to relate frictional resistance of two objects rubbing on each other, determined by dividing the frictional force by the compressive load across the bearing.

Component of a force Portion of a force acting in a particular direction or directions.

Components of force Breakdown of a force in different directions. The vector sum of all force components is just equal to the original force. Thus deformation and reaction to force components are the same as to the original force.

Compression Application of force tending to squeeze or crush an object.

Corrosion Destruction of metal artifacts by electrochemical action.

Critical load Vertical force that begins to produce buckling.

Deceleration Rate at which a moving object is slowed (same units as acceleration).

Ductile May be deformed permanently without fracture (ie, can be drawn into a wire or rolled into a sheet).

Elastic Deformation that disappears when the stress is removed.

Elastic (Young's) modulus Measure of relative stiffness. It is determined by dividing the stress (newtons/m^2) by the strain (%) and therefore has the same units as the stress. May also be thought of as the proportionality constant relating a material's stress and strain behavior.

Elasticity Deformation of an object when the stress depends only on the magnitude of the strain, independent of the rate at which the object is being strained or deformed. When the stress is removed, the strain disappears.

Elastohydrodynamic phenomenon The friction-lowering advantage obtained when the bearing surfaces are elastic in nature.

Elongation at fracture Permanent (percentage) deformation remaining at fracture; ductile material has a larger elongation at fracture than does brittle material.

Extreme fibers Outermost fibers on the convex and concave sides of a bent object.

Fatigue fracture Structural failure caused by repetitive tensile stresses that, although below the ultimate strength, cause a slowly propagating crack to cross the material.

Fatigue limit Repetitive stress that can be endured indefinitely by a particular metal; for stresses below the fatigue limit, fatigue life is infinite.

Forging Fabrication by mechanical deformation.

Fracture Failure caused by the growth of a crack.

Free body analysis Method of determining forces acting on a body by isolating that body and ensuring that it is in static equilibrium.

Hydrodynamic lubrication Situation in which the two bearing surfaces are separated by a fluid film held in place by the relative motion of the two bearing surfaces.

Hydrostatic lubrication Creation of a fluid film by pressurizing the fluid.

Hydrostatic pressure Stress produced by forces acting equally in all directions.

Inertia Tendency of a mass to resist changing velocity.

Kinetic energy Energy achieved by the motion of an object, determined by multiplying one-half the mass by the square of the velocity that the object is moving.

Mechanical energy Energy stored in the form of elastic stresses and strains; originally, this energy was work done by applied forces.

Metabolic energy Processes by which a living organism creates energy from its food.

Microklutz An individual with minor (momentary) incoordination.

Moment arm Shortest distance between the line of application of a force and the point of interest around which the moment of the force is being taken.

Momentum Mass of the object multiplied by its velocity.

Neutral axis Plane in a bent object at which zero stresses and strains occur.

Newton's first law If the sum of the forces on a stationary object is zero, the object does not move (or, if at all, continues to move at a constant velocity).

Newton's third law For each force there is an equal and opposite force.

Partial body mass The mass part of the body. For example, the mass acting on the hip joint on the swing side in gait is a partial body mass, as the weight of the swing leg distal to that hip is not being supported by that hip.

Piezoelectric Solid that responds to applied stresses by becoming electrically polarized. A voltage occurs when forces are applied.

Plastic flow Deformation caused by shear stress, resulting in permanent changes in the shape of any solid.

Poisson's ratio Ratio of the strain perpendicular to the line of force application divided by the strain parallel to the line of force application.

Polar moment of inertia Resistance of a given cross section to twisting; the stress caused by twisting (torsion) is proportional to the torque and inversely proportional to the polar moment of inertia.

Potential energy Energy created by work done against the force of gravity (ie, an increase in the height of the center of mass).

Power Work or energy per unit of time.

Resultant force Sum of force components.

Self-tapping machine screw A screw that cuts its own threads in the bone (or other medium) as it turns.

Shear Force applied parallel to an object's surface (eg, rubbing force). Shearing forces can also exist deep within the material itself.

Shear stress Force applied parallel to the surface that tends either to create friction (at the surface) or to shear deformation of the interior of the material.

Static equilibrium State at which the sum of the forces acting on a body is zero (Newton's first law is satisfied).

Stiffness Resistance to strain (deformation).

Strain Amount of deformation (percent elongation) compared with original dimension.

Strength Maximum resistance to stress before failure.

Stress Force per unit of area.

Stress concentration Point at which the stress is appreciably higher than elsewhere due to the geometry of the stressed object or the point of application of the force.

Surface energy Energy required to create new surfaces; essentially the energy of broken chemical bonds.

Tensile stress Tensile (stretching) force divided by the cross-sectional area.

Tension Application of force tending to elongate an object (a pull).

Tension band Member that is put in tension in order to compress other portions of the structure (eg, a guy wire).

Thixotrophy Viscosity that is shear rate dependent.

Torque Twisting moment (force times lever arm).

Torsion Forces applied tending to rotate an object about its long axis (a twist).

Toughness Energy necessary for fracture. A soft, ductile material may be relatively tough.

Ultimate strength Stress at which material ruptures.

Ultimate tensile stress Maximum tensile stress sustainable by a given material.

Vector Graphic representation of a force as an arrow. The direction of the arrow is the line of action of the force; the length of the arrow is proportional to the magnitude of the force.

Viscoelasticity When stress depends on the rate of strain as well as on the magnitude of the strain.

Weeping lubrication Special form of hydrostatic lubrication in which the interstitial fluid of cartilage is pressed out into the joint space by the deformation of cartilage under load.

Work Force times distance.

Yield strength Stress necessary to cause plastic flow (also known as yield stress)

Index

Page numbers followed by *f* indicate figures, and those followed by *t* indicate tables.

A

Abrasive wear, 196, 196*f*
Acceleration, 109
Acetabular prosthesis, stress distribution following hip replacement, 174–176, 175*f*, 176*f*
   at cement-bone interface, 176–178, 177*f*, 178*f*
Adhesive wear, 196–197, 196*f*, 197*f*
Anisotropic materials, tensile stresses in, 60
Anulus fibrosus, hydrostatic pressure and, 21, 21*f*, 22*f*
Applied force
   fracture mechanics and, 63
   versus stresses developed, 13–15
Area moment of inertia, 58
Articular cartilage
   fibrocartilaginous healing and, 151–152, 152*f*
   and frictional resistance, reducing mechanisms, 152–158, 154*f*–157*f*
   mechanical behavior of, 138
   stress distribution role of, 136–138, 136*f*, 137*f*
   wearing away of, 139–143, 139*f*, 140*f*, 142*f*, 143*f*. *See also* Osteoarthrosis
Articulations, in long bones, 56, 56*f*

B

Baseball, catching injuries and techniques, 116–118, 116*f*–118*f*
Bending
   in different materials, 60–61, 60*f*, 61*f*
   long bones, 54–59, 54*f*–58*f*
   of spinal column. *See also* Scoliosis
      lateral, braces preventing, 26*f*, 26–27
      resistance to, 28–31, 29*f*–31*f*
      role in, 15–22, 16*f*–18*f*, 20*f*
Bending moments, 33, 33*f*–35*f*, 35–37, 36*f*, 37*f*
   correcting scoliosis, 37–44, 38*f*–44*f*
Biaxial muscles, stress reduction and, 57, 57*f*
Body jacket, 42*f*
Bone(s)
   and cementless prosthesis ingrowth, 80, 81*f*
   fractures, conditions causing, 5
   long. *See* Long bones
   processes in, during loading, 53
   ultrastructure of, mechanical properties and, 69
   Young's modulus of, 80*t*
Bone cement, for rigid fixation of prosthesis, 160*f*, 160–161
Bone grafting, long bones, 104, 105*f*–106*f*, 106–107
Bone necrosis, corrosion products causing, 70
Boundary lubrication, 153, 154*f*
Braces, 38–43, 41*f*–45*f*
   of lumbosacral spine, 25–27, 26*f*
Brittle materials
   energetics of fracture in, 64–65, 64*f*
   fracture of, relation for, 63
   tensile stresses in, 61, 63

Buckling force, 28–31, 29*f*
Bursitis, 131, 132*f*

C

Cane, in osteoarthrosis treatment, 144–146, 146*f*
Cartilage. *See* Articular cartilage
Casts, 37–38, 39*f*, 40*f*
Center of gravity, locomotion and, 125–126, 127*f*–128*f*, 128
"Charley horse," 131
Clover leaf rod, 101*f*, 103
Cobalt-chromium-molybdenum alloys, as implant materials, 77–79, 78*t*
Cold flow, in polyethylene. *See* Polyethylene
Collagen
 articular cartilage and, 140*f*, 140–141
 in bone, 69
Compression
 shear force converted to, 14*f*, 14–15
 of spinal column, 2–4, 3*f*
 mechanics, 8, 8*f*
Compression fractures
 spinal column, 4*f*, 4–5, 5*f*
 sports-related, 113, 113*f*
Compression plate fixation, 88–89
Compression rods, 45–46, 46*f*–48*f*
Corrosion, of metallic implants, 70–73, 70*f*, 71*f*
Cotrel-Dubosset system, 50, 51*f*
Creep, in polyethylene. *See* Polyethylene
Crevice corrosion, 72*f*, 72–73, 73*f*
Crutches, in osteoarthrosis treatment, 144*f*, 144–146, 145*f*
Curvature, of the spine. *See* Scoliosis

D

Deceleration process, 110
 baseball catchers and, 116*f*, 116–117, 117*f*–118*f*
 football tackling and, 112*f*, 112–114
 and shock absorption, 111, 111*f*
 slowing of, 111–112
 untimely, 115*f*, 116
Derotation systems, scoliosis correction by, 50, 51*f*
Deyerle pin and plate, 99, 99*f*
Disc(s)
 bone versus, stiffness of, 3–4
 degeneration of, stress concentration and, 22–23, 23*f*
 role of, 22–23
 spinal compression and, 5, 5*f*
Distraction rods, 44–46, 46*f*–48*f*
Ductile materials
 energetics of fracture in, 64–65, 65*f*
 tensile stresses in, 61
Dwyer apparatus, 49*f*, 49–50

E

Elastic modulus, spinal compression and, 4, 5
Elastohydrodynamic phenomenon, 155–156, 156*f*
Electrical activity, bone mechanics and, 82–83, 83*f*
Energy
 dissipation of. *See* Shock absorption
 kinetic. *See* Kinetic energy
 mechanical, fracture and, 59
 metabolic, 109, 128
 potential, in locomotion, 124, 125–126, 125*f*, 126*f*
 surface (tension), 63
Exercises, lumbosacral flexion, 25, 26*f*
Extension injuries, sports-related, 113–114, 114*f*
External fixation, for fractures, 103–104, 104*f*
Extreme fibers, bending and, 17, 17*f*, 58

F

Fatigue cracks, 65*f*–66*f*, 66, 197, 198*f*
Fatigue failure. *See also* Fatigue cracks; Fatigue fracture

loading cycle data and, 67f
macrostructure of, 68f, 68–69
in muscles, 131
sports injuries and, 120–124
Fatigue fracture, 65–69
spondylolisthesis, 27f, 27–28, 28f
Femoral neck fracture, internal fixation for, 96–101, 96f–103f
Femoral prosthesis
cement and bone surrounding, forces and stresses in, 183–184
stress distribution in, 178, 179f–180f, 181
secondary to fixation, 181–182, 182f
Fibroblasts, mechanical stress effects on, 83–84, 84f
Fibrocartilaginous healing, mechanical stimulation of, 151–152, 152f
Fixation. See Internal fixation
Flexibility, of trunk, 1, 2f
Flexion exercises, 25, 26f
Flexion injuries, sports-related, 113–114, 114f
Football, injury types, 112–114
Force(s)
applied versus stresses developed, 13–15, 13f, 14f
components of, 37, 37f
compression and, 2–3
in normal hip joint, 161–170, 162f-170f
spinal buckling and, 28–31, 29f, 30f
straightening a curved spine, 31–37, 31f–37f
vectors, 10–13, 11f–12f
Forearm, reducing stress in, 122, 123f
Fracture(s). See also Bone(s)
causes of, 5
compression type, spinal column, 4f, 4–5, 5f
cortical bone resistance to, 65–69
energetics of, fracture toughness, and impact, 64–65
fatigue type, 65–69
spondylolisthesis as, 27f, 27–28, 28f
"march" type, 68f, 69, 69f
mechanics of, 53

in long bones, 54–59
tensile stress and stress concentrations in, 59–64, 59f–64f
spiral type. See Spiral fractures
treatment and repair of
fixation devices. See under Internal fixation
mechanical considerations, 82–85
Fracture toughness, 64f, 64–65
Free body analysis, 36
Frictional force, 10, 10f
Frictional resistance
between joint and cartilage, 153
lubricating mechanisms to reduce, 153–158, 154f–157f
prosthetic acetabular stress distribution and, 174–176, 175f, 176f
Fusion, of spinal segments, stress concentration and, 22–23, 23f

G

Gait cycle, phases in, 128, 129f–130f, 130
Gravity, center of, locomotion and, 125–126, 126f–128f, 128
"Greenstick" fracture, 53

H

Halo-pelvic traction, 37, 38f
Hard material. See Brittle materials
Harrington instrumentation, scoliosis and, 44–49, 44f–49f
Haversian canals, fatigue failure and, 68f, 69, 69f
HDPE. See Polyethylene
Hip joint
normal, forces in, 161–170, 161f–170f
osteoarthrosis of, 144–149
treatments for, 147, 148f–150f, 150
stress distribution following replacement, 171–174
acetabular component. See Acetabular prosthesis

Hip joint *(Continued)*
    femoral component. *See* Femoral prosthesis
    wear mechanisms following replacement, 197
Humerus, reducing stress in, 122, 123*f*, 124
Hydrodynamic lubrication, 154, 155*f*
Hydrostatic lubrication, 155, 156*f*
Hydrostatic pressure, bending and, 20*f*–21*f*, 21, 22*f*

I

Iliotibial band, tensile stress reduction and, 121–122, 122*f*
Ilizarov technique, 104*f*
Implant materials
    cobalt-chromium-molybdenum alloys, 77–79
    interaction with bone, 79–81
    mechanical properties of, 78*t*
    polymers, 81–82
    selection criteria, 74
    stainless steel, 75, 77
    tensile properties of, 74, 75*f*, 76*f*
    titanium and titanium "six-four" alloy, 79
    Young's modulus of, 80*t*
Impulsive loading
    deceleration failure and, 115*f*, 116
    protection against, 119*f*, 119–120
Inertia
    area moment of, 58
    polar moment of, 59
Internal fixation
    of fractures, 87–90, 87*f*–90*f*
        bone grafting, 104–107, 105*f*, 106*f*
        devices. *See under* Internal fixation devices
    rigid, of prosthesis, 160*f*, 160–161
Internal fixation devices
    fractures
        implant materials used, 74–82, 75*f*, 76*f*, 78*f*, 80*f*–82*f*. *See also* Implant materials; *specific materials*
    metallic implants, corrosion of, 70–73
    nails, rods, and pins, 96–104
    plates, 87–90, 87*f*–91*f*
    screws, 93*f*, 93–94, 94*f*–95*f*, 96, 96*f*–103*f*
    wire and tension bands, 85–87, 85*f*–87*f*
    spinal column curvature
        derotation systems, 50, 51*f*
        Dwyer apparatus, 49–50, 49*f*
        Harrington instrumentation, 44–49, 46*f*, 47*f*
Intertrochanteric fracture, stabilization of, 100*f*–101*f*, 101–102
Intramedullary rods, for fracture fixation, 101*f*–103*f*, 102–103
    spiral fractures, 92*f*, 93

J

Joint(s)
    articular cartilage and. *See* Articular cartilage
    decreasing load on, 144–149, 144*f*–149*f*
    degeneration of. *See* Articular cartilage; Osteoarthrosis
    hip. *See* Hip joint
    increasing weight-bearing area of, 149–151, 149*f*, 150*f*
    knee. *See* Knee joint
    lubricating mechanisms
        artificial, 158
        natural, 152–158, 154*f*–157*f*
    replacement of, 159–161, 160*f*
    stress distribution within, 136*f*, 136–138, 137*f*
    wear mechanisms following replacement, 196–199, 196*f*–198*f*

K

Kinetic energy, 109
    in locomotion, 125*f*, 125–126, 127*f*–128*f*, 128

Knee joint
  deterioration effects, 188
  patello-femoral component, 194–196, 195f
  prosthesis design and materials, 188–196, 189f–195f
  replacement considerations, 191–196, 192f–196f
  stability of, factors determining, 184–188, 185f–188f
  wear mechanisms following replacement, 196–199
Kuntschner rod, 101f, 103
Kyphosis, Milwaukee brace correcting, 42–43, 45f

L

Loading
  eccentric, bending of spine and, 15, 16f
  fatigue test data, 7f
  processes in bone during, 53
Loading patterns
  articular cartilage fibrillation and, 143, 143f
  bone fractures and, 5, 5f
Localizer jacket, 37, 38, 40f
Locomotion, mechanics of, 124–130. *See also* Running
Long bones
  bone grafting and, 104, 105f–106f, 106–107
  as levers, 58, 59f
  strain gauges and, 58–59
  structure of
    circulation, repair and nutrition, 58
    stress reduction and, 57, 57f
    tensile stresses in, 54–59, 54f–57f, 59f
Lubrication, of joints
  artificial, in osteoarthrosis, 158
  natural mechanisms, 152–158, 154f–157f
Lumbosacral spine
  bending and, 17–18, 18f
  flexion exercises and bracing, 25, 26f

M

"March" fracture, 69. *See also* Fatigue failure; Fatigue fracture
Mechanical aids, in osteoarthrosis treatment, 144–146, 144f–146f
Mechanical energy, fracture and, 59
Meniscal tear, 132f, 133
Metabolic energy, 109, 128
Metallic implants, corrosion of, 70–73, 70f–73f
Microfracture, shock absorption and, 124
Microklutzes, 115f, 116
Milwaukee brace, 38–41, 41f–43f
  correcting kyphotic deformity, 42–43, 45f
Mineral, in bone, 69
Momentum, 109
MP35N, as implant material, 78
Multiple-screw plate fixation, 87–88, 87f–88f
  spiral fracture, 92f, 93
Muscle(s)
  rupture, 130–131, 131f
  sports injuries and, 120–124, 121f
  stress reduction and, 56–57, 57f
  tear, 131

N

Nail fixation, 96–104, 99f
Neutral axis, bending and, 15, 16f, 17
Newton's first law, spinal column mechanics, 6
Newton's third law
  spinal column mechanics, 7
  sports injuries and, 109–118
Norton-Brown brace, 25
Nucleus pulposus, hydrostatic pressure and, 21, 22f

O

Olecranon process, wire fixation of, 86, 86f
Osteoarthrosis, 135
   articular cartilage thinning in, 141, 142f
   treatment of, 143–144
      artificial lubricants, 158
      decreasing load on joint, 144–149, 144f–146f
      increasing weight-bearing area of joint, 149–151, 149f, 150f
      mechanical stimulation, 151–152, 152f
Osteopenia, implant materials and, 79–80, 81f
Osteotomy, in osteoarthrosis treatment
   valgus, 149, 149f, 150f, 151
   varus, 147–148, 148f, 149f–150f, 150
Osteotomy compression hook, 100f,

P

Pars interarticularis, fracture of, 27f, 27–28, 28f
Pelvis, obliquities of, 25–27, 26f
Pin fixation, for fractures, 96–104
   Sage pin, 103
Plastic flow, 53
Plate fixation
   compression type, 88–89
   multiple-screw, 87–88, 87f–88f
   single versus double, 89, 89f, 90f
   for spiral fractures, 90–93, 91f–93f
Poisson's ratio, 8
   tension and, 9
Polar moment of inertia, 59
Polyethylene
   hip replacement and, 175–176, 176f
   knee replacement and, 194–196, 195f
   wear mechanisms and, 197, 198f, 198–199
Polymers, as implant materials, 81–82, 82t. *See also* Polyethylene
Potential energy, in locomotion, 124, 125–126, 126f
Prosthesis(es)
   acetabular. *See* Acetabular prosthesis,
   cementless, ingrowth of, 80, 81f
   femoral. *See* Femoral prosthesis
   material used in. *See* Polyethylene
   replacing joints, 159–161, 160f. *See also* Hip joint; Knee joint
   wear associated with, 196–199, 196f–198f

R

Reaction force, 110, 110f
Replacement, of damaged joint. *See* Hip joint; Knee joint; Prosthesis(es)
Rod fixation
   for fractures, 101f–103f, 102–103
   spinal curvature and, 44–51, 46f–49f
   spiral fracture and, 92f, 93
Running
   double support phase and, 130
   mechanics of, 124–130
   reaction energy and, 110, 110f
Rupture, muscle, 130–131, 131f
Rush rod, 103
"Rust granulomas," corrosion products causing, 70

S

Sage pin, 103
Scoliosis
   forces correcting, 31, 31f
   internal fixation devices for, 44–51, 46f–49f
   spinal rotation and, 29–31, 30f
   straightening curvature in
      devices used, 37–44, 38f–45f
      mechanics of, 31–37, 32f–37f
Screw fixation, 93–94, 94f
   spiral fracture, 93, 93f
   thread design, 94, 95f, 96
   wire fixation versus, 87, 87f
"Self-tapping" machine screw, 94
Shear force, 9–10, 10f

in acetabular component of replaced hip joint, 174–176, 175f–176f
conversion to compression, 14f, 14–15
in forward movement of spine, 14, 14f
Shear strains, 60, 60f
Shear stresses, prosthesis transmission of, 160f, 160–161
Shock absorption
　active musculoskeletal, 119f, 119–120
　effectiveness of, 110–111, 111f
　equipment and techniques to enhance, 118
　fatigue and. See Fatigue failure
　reaction force and, 110
Single plate fixation, 89, 89f
Skis, shock absorption and, 118
Snowmobiles, shock absorption and, 118
Soft material. See Ductile materials
Soft tissue injury, 132f, 133
Spinal column
　and bending. See Bending
　compression and compression fractures, 2–5, 3f–5f
　curvature of. See Scoliosis
　curvature of, straightening mechanics, 31–37, 32f–37f
　disc role, 22–23
　forces applied versus stresses developed, 13f–14f, 13–15
　functional anatomy, 1–2, 2f
　internal fixation devices, 44–51, 46f–49f
　lumbosacral flexion exercises and spinal bracing, 25–27, 26f
　mechanics of, 6–8, 6f–8f
　shear force, 9–10, 10f
　spondylolisthesis, 27f, 27–28, 28f
　tension, 8–9, 9f
　torsion, 24, 24f–25f
　traction, casts, and braces, 37–44, 38f–45f
　vectors, 10–13, 11f–12f
Spinal fusion, stress concentration and, 22–23, 23f
Spiral fractures, 60, 60f, 90–93, 91f–93f
　plate fixation of, 90–91, 90f–92f, 93
　screw fixation of, 93, 93f
Spondylolisthesis, as fatigue fracture, 27f, 27–28, 28f
Sports injuries
　from fatigue, 120–124, 121f–124f
　forms of, 130–133, 131f–133f
　locomotion mechanics, 124–130, 125f–130f
　musculoskeletal shock absorption, 119–120, 119f
　and Newton's third law, 109–118, 110f–118f
Stainless steel, as implant material, 75, 77
Stellite 21, as implant material, 77
Stellite 25, as implant material, 78
Stiffness
　bending mechanisms and, 19
　bone versus disc, 3–4
　of materials, 4–5
Stiffness gradients, cartilage fibrillation and, 143, 143f
Strain(s), spinal compression and, 3. See also Bending
　measurement of, 4
　on vertebral body-disc unit, 5, 5f
Strength, of materials, 4–5
Stress(es)
　developed, forces applied versus, 13–15, 13f–14f
　distribution within joints, 136f, 136–138, 137f
　fatigue limit, 66, 68, 67f–68f
　spinal compression and, 3. See also Bending
　measurement of, 3
　torsion, 24, 24f–25f
Stress concentration
　disc degeneration producing, 22, 23f
　spinal fusion producing, 22–23f
　tensile stresses and, 61f, 61–63, 62f
Stress fracture. See Fatigue failure; Fatigue fracture
Surface energy/tension, 63
Synovial fluid, as thixotropic fluid, 157f, 157–158
Synovial joint, friction under different loads, 156, 157f

T

Tear(s)
  meniscal, 132f, 133
  muscle, 131
Tendonitis, 131, 132f, 133
Tennis elbow, 131, 133
Tenotomy, in osteoarthrosis treatment, 146, 147f
Tensile strength
  measurement of, 9
  of polymers, 82t
Tensile stress, 8–9, 9f
  in articular cartilage, 139, 139f
  in long bones, 54–59, 54f–57f, 59f
  perpendicular verus horizontal, 63–64, 64f
  reducing. See Tension band
Tension band fixation, 85–87, 85f–87f
  iliotibial, 121–122, 122f
  by osteotomy compression hook, 100f, 101
  and pin placement, 97f, 98
Texas Scottish Rite Hospital system, 50, 51f
Third-body wear, 197, 198f
Thixotropy, 157f, 157–158
Thoracal lumbosacral orthosis (TLSO), 42–43, 44f
Thread design, screw fixation and, 94, 95f, 96
Titanium, as implant material, 79
Torsion stresses
  long bones, 54–59
  spinal column and, 24, 24f–25f
Traction, 37, 38f
Trunk, flexibility of, 1, 2f

Turnbuckle jacket, 37, 39f
Two-plate fixation, 89, 90f

V

Vectors, 10–13, 11f–12f
Vertebrae
  compression and compression fractures, 2–5
  Dwyer method and, 49f, 49–50
  torsion transmission in, 24, 25f
Viscoelasticity, 138
Viscous flow, 53

W

Walkers, in osteoarthrosis treatment, 144–146
Wear
  abrasive, 196, 196f
  adhesive, 196–197, 197f
  fatigue cracks, 197, 198f
  following joint replacement, 196–199
  osteoarthrosis. See Osteoarthrosis
  third-body type, 197, 198f
Weeping lubrication, 154, 155f
Williams Exercise Program, 25, 26f
Wire fixation, 85–87

Y

Young's modulus
  of bone and implant materials, 80t, 82t
  spinal compression and, 4, 5